THE *fresh*
HONEY
COOKBOOK

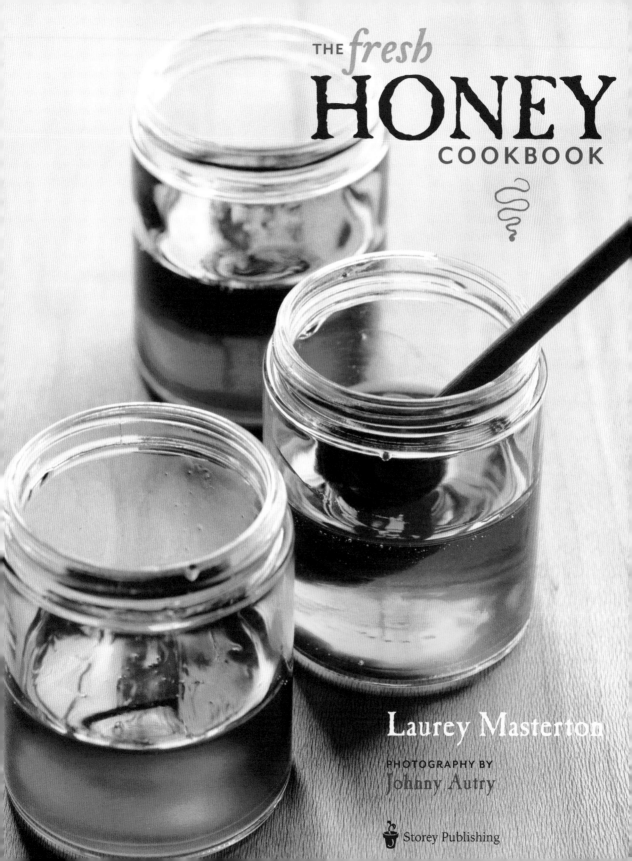

THE *fresh* HONEY COOKBOOK

Laurey Masterton

PHOTOGRAPHY BY
Johnny Autry

Storey Publishing

The mission of Storey Publishing is to serve our customers by publishing practical information that encourages personal independence in harmony with the environment.

Edited by
Margaret Sutherland and Sarah Guare
Art direction and book design by
Carolyn Eckert
Text production by Theresa Wiscovitch

Photography by © Johnny Autry
Photo styling by Charlotte Autry
Beehive illustration, page 41, by © the author

Indexed by Christine R. Lindemer,
Boston Road Communications

The information in this book is true and complete to the best of our knowledge. All recommendations are made without guarantee on the part of the author or Storey Publishing. The author and publisher disclaim any liability in connection with the use of this information.

Storey books are available for special premium and promotional uses and for customized editions. For further information, please call 1-800-793-9396.

Storey Publishing
210 MASS MoCA Way
North Adams, MA 01247
www.storey.com

Printed in China by Toppan Leefung Printing Ltd.
10 9 8 7 6 5 4 3 2 1

Library of Congress Cataloging-in-Publication Data

Masterton, Laurey.
  The fresh honey cookbook / by Laurey Masterton.
    pages cm
  Includes index.
  Includes bibliographical references and
    index.
  ISBN 978-1-61212-051-5 (pbk. : alk.
    paper)
  ISBN 978-1-60342-871-2 (ebook)
  1. Cooking (Honey) I. Title.
TX767.H7M38 2013
641.6'8—dc23

                                    2012047104

Storey Publishing is committed to making environmentally responsible manufacturing decisions. This book was printed in the United States on paper made from sustainably harvested fiber.

dedication

To my mentors, Carl and Debra, who are teaching me how to dance with the bees

To my wonderful sisters, Lucinda and Heather, who love and support me in all things

To my sweet, sweet Barbee, who inspired me to start learning about bees

And most of all, to the bees, the most inspirational teachers of all

thanks

To Lisa and Sally Ekus, the most supportive, loving, and hard-working agents I could ever imagine. Thank you so much for everything. I mean everything!

To Margaret Sutherland, Sarah Guare, Carolyn Eckert, and everyone at Storey Publishing for shepherding this book from the inside of me out into the world.

To Dianne Tuttle, for amazing support and love and for being such a fine-toothed proofreader.

To Charlotte and Johnny Autry for photographic and food-styling magnificence.

To my recipe testers: Kim Austin, Jan Brunk, Drew Gladding, Barbara Hammer, Chet Holden, Marlisa Mills, Eleanor Owen, Cindy Platt, Kim Rosenstein, Blake Swihart, Jane Ann and Phineas Tager, Adam and Emily Thome, and Noel Weber.

To the entire gang at Laurey's: Emily, Adam, Noel, Leslie, Brendan, Lito, Rolando, Martha, Deb, Marty, Barbara,

Edith, Andrew, Andy B., Andy L., Jason, Ari, Evelyn, Austin, Irvegg, and especially to my favorite hero of all, Henry.

To my doctors, without whom I truly would not be here: Dr. Paul Ahearne and Amy Antczak; Dr. Michael Messino and his incredible team: Tina Messer, Debbie Splain; and especially Charlotte Lail, Debbie Payne, and Janet Magruder, my true Guardian Angels; and Dr. Benjamin Calvo and Teresa Sadiq.

To the many, many hundreds of prayers and offerings and thoughts of love and light that lifted me, held me, and carried me as I waded through a year of treatment, all while writing this book.

To Livestrong at the YMCA of Asheville and to Livestrong and the Lance Armstrong Foundation for helping me get back on my feet.

Gracias Recibidas

Thanks to St. Peregrine and to the heart Milagro for prayers received.

And finally, thanks to Alicia.

contents

winter

spring

preface

I grew up around food. My parents, Elsie and John Masterton, founded Blueberry Hill Inn in Goshen, Vermont. I grew up there and loved helping my mother cook and my father host guests in the inn. My mother wrote the Blueberry Hill series of cookbooks, which got me started as a cook. My first solo cooking triumph, at the age of six, was the successful completion of a batch of my mother's brownies.

Though I had planned to run Blueberry Hill Inn when I grew up, my parents' deaths when I was 12 sent me on a different path. My two sisters and I tried to live with other families and, when that didn't work, went away to boarding schools and then college. Each move took us farther from Blueberry Hill. Life with my beloved inn seemed impossible, so I set my sights elsewhere and pursued various side routes, following interesting invitations and detours. I worked on a fishing boat and repaired fishing nets. I washed dishes in a restaurant and at a summer camp. I designed stores and commercial showrooms. I worked as a theatrical designer in Off-Broadway, Off-Off-Broadway, and Broadway theaters in New York City, and finally, searching for a way to get out of dark theaters, I made my way to North Carolina, where I attended an Outward Bound course and later became an instructor.

Food was never far from my heart. And though running Blueberry Hill was not a possibility for me, I realized that I could still cook for my livelihood. I plunged in, launching a catering company from my tiny Asheville apartment in 1987. I managed to cobble together enough work so that by 1990 I was actually supporting myself. The local health department got wind of my venture one fateful day, however, and I was forced to either quit or grow, which I did (grow, that is), moving to the sleepy downtown of Asheville well before its renaissance. Originally operating just a catering kitchen, I gave in to customer pressure and expanded from 2 to 14 seats, adding retail and a café.

Laurey's is now in its fourth location and comprises a 50-seat café and a busy catering company with a full staff of what I call "talented and interesting individuals": artists, musicians, and creative people of all kinds. We serve "gourmet comfort food" in an airy old horse-drawn-carriage-making building, just a block away from the heart of Asheville. We get our food from as many local sources as we can, totaling as many as 30 different farmers and local beekeepers when our growing season is at its height. A big part of my vision, along with making great food, is "to take care of the Earth," a point that drives me and informs the direction and mission of my business.

a sweet education

A few years ago, I was invited to cater a party for The Honeybee Project, an Asheville-based group that teaches children about the importance of honeybees to our food supply. After talking with the party's host, I decided to make only foods that would not exist without honeybees. As I explored and researched the menu, I was amazed to learn that without honeybees we would not have nuts, avocados, strawberries, melons, apples, and many, many other foods.

After getting this glimpse, I wanted to learn more. Dave, one of my local honey suppliers, suggested I go to the local "bee school," and in 2007 I signed up to attend the introduction to beekeeping course to be held the following January.

Bee school thrilled me. By the end of the first morning of class, I had bought a bee jacket, thereby committing myself to jumping in even though I had very little idea of what I was undertaking. By the end of the first weekend, I had made a list of equipment I would need to get started. By the end of school, I had placed an order and paid for two colonies of bees, which would arrive as soon as the temperatures around Asheville got warm enough. I was about to become a beekeeper.

One day shortly after bee school ended, I offered to teach a class of schoolchildren about bees and honey in my shop. I brought in all my bee gear, reviewed my notes, and prepared to teach. After my presentation, I offered tastes of honey and some recipes I had prepared using foods that would not exist without honeybees.

All of the children enthusiastically dove into the strawberries and avocados, chanting "One! Two! *Three!* This bite is the third bite!" I looked over at Susan, their teacher, who was sitting with some of her students. What was this all about?

"The children have been studying," she said, "and they know that every third bite they eat would not exist if it weren't for honeybees."

Every third bite? Really? I had never heard this before. Every. Third. Bite. Wow. Where would we be without honeybees? Where, for that matter, would I, a restaurant owner, be without food to cook? I knew bees made honey, but no one *depends* on honey, even though we may like it. But *ingredients*, a third of all we eat? Now that caught my attention.

I started to take this more seriously.

the start of a hobby

In February, just a few weeks after completing bee school, my equipment arrived. I hauled boxes of beehive parts home and turned my garage into "Bee World," assembling beehives and honey frames. In late April, I got a call saying my bees were ready. One warm afternoon a couple of days later, my sister and a friend stood by and took pictures of me cautiously taking my very first frames of live bees out of their travel boxes, and tucking them into their new beehive homes that I had built in my garage. I put those hives on a shady hill behind my house. I visited them occasionally, but mostly left them alone, figuring that nature was smarter than I and would take care of them. It was exciting to have bees, and I assumed that all would be well. I assumed, cockily, that I was a successful beekeeper.

I was wrong.

At the end of that first year of beekeeping, I discovered that all my bees had died. I went into the winter with empty hives. We'd had a drought that summer, however, and many people had bee losses. At my beekeepers club meetings, I heard talk of a limited sourwood honey flow. I assumed that I was just like everyone around here, losing bees. Not great news, but nothing too out of the ordinary.

Not giving up, I ordered more bees. However, I had planned an extended bike ride that would take me across the entire United States the following spring and so arranged to have someone else install my bees in the hives. When I returned from my bike ride, I continued to keep a distance from my bees, visiting them occasionally. I still felt like a beekeeper, but one with less confidence. And I sure did not feel that I was a successful beekeeper. At the end of that summer, I enlisted the help of a friend, who noted that my colonies did not seem very strong. She was not sure if my bees would last through the winter. By the time we looked, in late October, I realized that I was going to probably lose my bees for a second time. And that's exactly what happened.

In the beginning of the next year, I took my beehives filled with dead bees to my local beekeepers' group to see if anyone could explain to me what had happened. I did not think I had done anything deliberately wrong, but I was confused. And having dead bees was not really what I had in mind when I imagined myself as a beekeeper. Instead of getting sympathy, however, I got a sharp awakening to the realization that I would need to be much more involved if I was going to be a successful beekeeper.

"You can't just leave them alone, you know," one fellow scolded. "You need to make sure they are okay, and if they are not, you need to help them. They are living beings, not lawn ornaments!"

I studied more and, for starters, discovered that bees like sun, not shade. This meant that my bees would do much better in my front yard than hidden away in the back. I found a mentor who coached me, helped me learn how and when to visit my bees, and taught me how to assess the colony's strength and what to do if something was wrong. I studied for and passed the Certified Beekeeper's exam. And I pledged to learn and do even more.

a life shift

My bees are now prominently placed in the middle of my front yard. They are also in the center of my life and in the photographs in this book. I watch them and care for them and pay attention to the weather and to the flowers and to their world. I sit nearby, listening to and watching them, following their flights, and noting if they are coming in with pollen on their legs. I have become much more attentive, keeping track of their health and population size. I have removed all pesticides from my home and have planted new gardens, filling them with flowers and herbs that bees love. I am still a relatively new beekeeper, but I am no longer an inattentive one. My bees are doing well, and I am humbly grateful.

At the same time, the more I learn about bees, the more I use honey in my cooking. Thanks to my work with the National Honey Board, I've learned about the many varietals of honey that exist in the United States and now search out honeys on my travels, both domestically and abroad. Traveling friends bring home exotic honeys, which makes me very happy. As a result, my honey palate is growing far more educated.

I love tasting the difference between dark avocado honey from California and light acacia honey from Tuscany. I inhale orange blossom honey, breathing in the essence of those delicate citrus trees. I add sage honey to my lemonade, infusing it with my own sprigs of the same herb from my garden beds. The resulting beverage is so much more distinctive, so much more exotic than it would be with simple white sugar. I am appreciative of the subtle and overt tastes that honeys impart and love playing with old recipes, changing the sweetener from sugar to honey, trying to match the varietal with the flavor profile of the recipe.

In this book I share with you the art, the disappointments, and the thrills I have found by adding bees and honey to my life. Bees, to me, are miraculous.

I've included seasonal recipes for the whole year, featuring honey varietals from around the world (don't worry; I'll tell you where to find them in the United States). I also share with you little stories, my glimpses into the world of the bees — how they do what they do. Maybe you'll be inspired to keep bees too. But even if not, you can still enjoy their honey. I trust you will also be much more appreciative of all they add to our world and aware of what you can do to make sure they continue to thrive.

Enjoy!

Lauren

a note on the recipes

Some of the recipes here began their lives in the Blueberry Hill series of cookbooks, written by my mother, Elsie Masterton. I feel that these recipes have become my own after cooking them for myself and my friends and in my café. Over time I have changed them to accommodate the more modern preference for less fat and sugar, changes that my mother probably would have made if she had lived, and by replacing sugar with honey in many of the recipes. The result is a much more interesting flavor, since honey — especially the more unusual varietals — has a far more complex flavor profile than sugar. I have also included many favorites from Laurey's, my café in Asheville, and, finally, some recipes shared by friends.

I have organized each chapter around a specific honey varietal, to acquaint you with the differences among varietals, but don't feel limited to using only that honey in a recipe. I make alternative suggestions in case you do not have the honey I recommend. If you'd like to pick up a particular varietal, however, I've listed where I found each of these honeys in the back of the book (see page 197).

As you will read in the season openers, the beekeeper has important tasks to do in each of the four seasons of the year. I describe those tasks and try to make sense of a complicated subject. I write from the perspective of where I live in the mountains of western North Carolina. Other places might have a milder or longer winter than I have, but seeing things from my perspective will, I hope, provide a good starting place for you to see the big picture.

It is also necessary to note that the recipes in this book are written from my geographical perspective, because I feature spring beets in March, for instance. You may be able to get spring beets in January where you live, or maybe not until June, or maybe you can find only what is in your grocery store's produce section. Not to worry. Feel free to adapt the recipes to meet your needs, making one of my spring dishes when the appropriate produce is ripe in your area. Of course, the fresher and more local the produce, the better, in my mind. Right out of the garden is best if you can plant your own. If you have a farmers' market or a farm stand nearby, that's good too. Here in Asheville, for instance, we have around 30 different farmers' markets each week of the growing season — an amazing bounty. Hmmm, maybe you should just come to Asheville! (If you do, come have lunch with me. I'd love to see you.)

One very important point is that not all of the recipes contain honey, but every recipe features ingredients that bees pollinate. Bees are responsible for pollinating a great deal of what we eat. Can you imagine living just on potatoes, wheat, and rice? Without bees, there would be no apples. No peaches. No berries. No guacamole, no fruit tarts, no citrus smoothies. A bleak picture, don't you think? Honey is a precious thing, and the beautifully special creatures that make it deserve to be placed at the top of a very big pedestal. Without them, our culinary life would be very bland indeed.

I am excited to introduce you to some of my favorite honeys, to share some of my favorite recipes, and to help you understand the world of bees. To be completely honest, I believe that the health of our Earth is in danger. But instead of being completely overwhelmed, I also believe that we can all do our part to make a difference. Taking care of the bees and appreciating honey are things that we can all do. They're small steps, but important ones.

Bee-Dependent Ingredients

In each recipe, items that honeybees produce or help produce, either directly (such as their honey) or as pollinators, are noted in **bold** type.

Chestnut Honey
Sage Honey
Avocado Honey
Raspberry Honey
Sourwood Honey
Cranberry Honey
Acacia Honey
Tulip Poplar Honey

how to taste honey

It can be incredibly daunting to be faced with a shelf full of honeys at a fine food store or a table full of varietals at the farmers' market. How do you decide which to get, assuming you are allowed to take a taste? I admit to once being as confused as you might be at the thought of tackling this complex subject. But, as with anything, taking a big subject and breaking it down into manageable parts is a good approach, and, with honey tasting, it really works.

Inspired by an article titled "A Taste of Honey" by Barbara Boyd, printed in the September 2011 issue of *Bee Culture* magazine, I now present honey tastings to interested groups of untrained honey tasters. By the end of the session, it is much easier to taste the subtleties in different varietals.

The tastings are broken into three parts: sight, smell, and taste.

sight

Is the honey liquid or solid? Remember that a crystallized honey simply needs to be warmed to return it to its liquid state, but it's good to know the difference. Also, once you become aware of the consistency of pure honey, you can easily recognize a jar of too thin liquid — an indication that you might be encountering an adulterated product from a questionable source.

Is the honey clear or are there impurities? Honey is usually filtered enough to remove little bee parts, but absolutely clear liquid is not necessarily a goal. If you can, talk to your beekeeper or honey salesperson to understand what you are seeing.

What color is it? Honey ranges in color from very light to very dark, from wheat to dark brown. It can be helpful to be able to clearly describe the color of your honey.

smell

How intense is the odor? You may like a strong aroma or you may prefer a milder one. Ask to smell the honey. One smell will tell you a lot. It is also easy to detect spoiled honey — it will smell fermented or rotten. And know that real, pure honey will never spoil. A rotten smell is a sure sign that someone has added sugar water or something else that does not belong.

How would you describe the smell? Many products, such as honey, wine, and olive oil, have a flavor wheel, which gives vocabulary for smelling and tasting. To use the honey flavor wheel on page 19, start in the center of the wheel and pick which of the seven descriptors most fits your perception of the aroma of the honey. Does it have a floral aroma? Is it a warm, vegetal, woody, chemical, fresh, or fruity smell? Or does it smell unpleasant, rotten, or spoiled? If you answered "yes" to the last question, this is not pure honey and you don't want it!

Each pie shape in the center of the wheel extends to more descriptions that will help define the smell for you. *Warm*, for instance, leads to *burned*, *cooked fruit*, *caramelized*, and *subtle*. Smell multiple times to become clearer on what you are smelling.

It is important to note that everyone's senses have subtle differences and, thus, no two people will have the same reaction to one honey. Some people love the strong, dark, sharp honeys. Others like light, mild ones. What's important in tasting is to educate your senses so you can understand what *you* like.

taste

Take a small taste of the honey by dipping a toothpick into the jar. If you need another taste, get another toothpick. You do not need to pour a puddle onto a plate!

Is the honey sweet or salty? Sour or bitter? We have taste receptors on different parts of our tongues, and once you are paying attention, you will notice things like a salty honey. Is the sweetness or bitterness mild or strong? Do you like what you are tasting?

Turning back to the flavor wheel, follow your second aroma descriptor out to the perimeter of the wheel. For instance, if you have decided after tasting that the honey smells *woody* and also *spicy*, you will find that your choice of taste descriptors are *clove*, *nutmeg*, and *coffee*. Which word most closely describes what you are tasting?

When tasting honey, pay attention to any aftertaste you might find. Some honeys are very mild and have no lingering flavor. Others can stay on the palate for a long time. You may or may not like that, but at least you'll know what to expect if you have tried the honey before buying it. If you like, keep your honey-tasting notes in a notebook. This could be especially helpful if, like me, you enjoy tasting honey on your travels and become overwhelmed trying to remember everything.

It takes some time to learn how to taste honey, but it does get easier with practice. I find that I am able to taste two or three honeys at most in one sitting. After that, no matter how much time I have taken between samples, my palate needs a rest. Over time, the more I have tasted, the more I can recall, and the more quickly I can evaluate the next honeys I try; this makes the process even more fun. Educating your palate to taste honey demystifies a complex subject, making it so much more enjoyable.

| avocado | chestnut | tupelo | tulip poplar | orange blossom | eucalyptus |

Honey Flavor Wheel

Beginning in the center of the wheel, choose the word that best describes the smell of the honey. Further refine this description by selecting a word in the next ring. Lastly, describe the honey's taste using a word in the outermost ring.

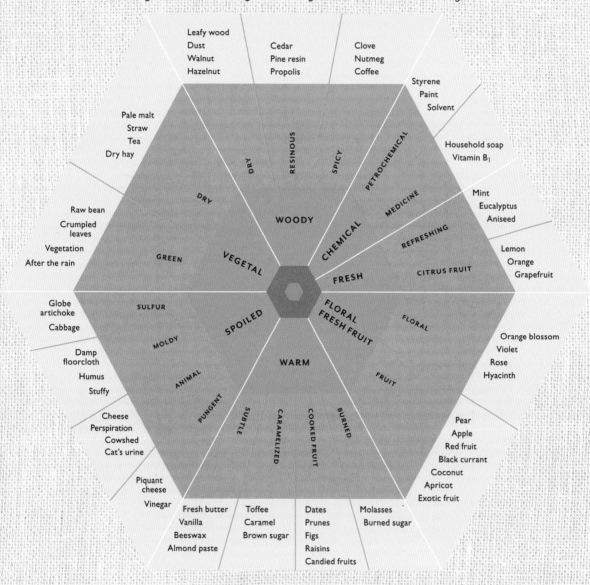

Leafy wood
Dust
Walnut
Hazelnut

Cedar
Pine resin
Propolis

Clove
Nutmeg
Coffee

Styrene
Paint
Solvent

Pale malt
Straw
Tea
Dry hay

Household soap
Vitamin B$_1$

DRY

RESINOUS

SPICY

PETROCHEMICAL

MEDICINE

Mint
Eucalyptus
Aniseed

Raw bean
Crumpled
leaves
Vegetation
After the rain

DRY

WOODY

CHEMICAL

REFRESHING

GREEN

VEGETAL

FRESH

CITRUS FRUIT

Lemon
Orange
Grapefruit

Globe
artichoke
Cabbage

SULFUR

SPOILED

FLORAL
FRESH FRUIT

FLORAL

Orange blossom
Violet
Rose
Hyacinth

Damp
floorcloth
Humus
Stuffy

MOLDY

WARM

FRUIT

Cheese
Perspiration
Cowshed
Cat's urine

ANIMAL

PUNGENT

SUBTLE

CARAMELIZED

COOKED FRUIT

BURNED

Pear
Apple
Red fruit
Black currant
Coconut
Apricot
Exotic fruit

Piquant
cheese
Vinegar

Fresh butter
Vanilla
Beeswax
Almond paste

Toffee
Caramel
Brown sugar

Dates
Prunes
Figs
Raisins
Candied fruits

Molasses
Burned sugar

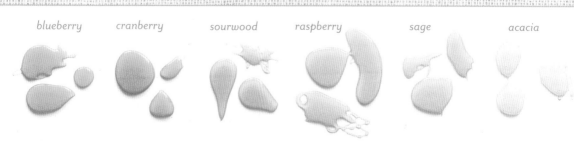

blueberry   *cranberry*   *sourwood*   *raspberry*   *sage*   *acacia*

winter
WHAT'S HAPPENING IN THE HIVE?

Where I live, the first part of the calendar year is a quiet time inside the hive. The colony is smaller than it is during the major honey production time. The queen remains tucked into the center of a cluster of bees, surrounded by workers who keep her and the entire hive at 95°F (35°C) during the chill of winter and the heat of summer. They are, as my mentor Debra tells me, "thermoregulating geniuses!" (more on that later).

During the winter, the queen is on hiatus from egg laying, and the workers live much longer than they do in the summer. They wear themselves out after about six weeks of busy activity during the summer, but in colder climates, they can live for up to six months.

As the temperature outside increases, the colony begins to get ready for the spring "honey flow," when the trees and plants bloom and have nectar available for the bees. The queen, knowing this, increases her egg production, setting in motion a huge change in the hive. In anticipation of these changes, the beekeeper needs to be ready too. It is time to go through the bee equipment, making sure that all is clean and in good repair.

On a warm day in mid- to late winter, the beekeeper can take a quick peek inside the hives. Ideally, it will be possible to see a healthy cluster of bees and plenty of honey in the frames. If the beekeeper doesn't find any honey, he or she will need to give the bees something to eat — preferably their own honey.

FEATURED HONEY VARIETAL:

orange blossom

There are certain pursuits which, if not wholly poetic and true, do at least suggest a nobler and finer relation to nature than we know. The keeping of bees, for instance. — HENRY DAVID THOREAU

Orange blossom honey is readily available due to the huge growing area of oranges. Spanning the southern United States from California to Florida, orange groves produce fruit in the early spring. Bees feast on the nectar from the fragrant white flowers (have you ever been in a blooming orange grove?), turning the sweetness into a light honey that pairs well with so many foods. I find that using it with recipes containing citrus fruits is a natural thing to do, the one easily complementing the other.

COLOR	SMELL	TASTE	AFTERTASTE
Light yellow/orange	Mild, freshly floral	Refreshing; mildly bitter with a medium sourness; citrus notes and orange rind	Mild, short-lasting

meyer lemon– and honey-marinated chicken skewers

If you can't find these lusciously different lemons, the regular kind will do, but Meyer lemons will give the chicken a distinctive and memorable flavor. Though this recipe started as a purely savory one, I think it works very well with a bit of honey. And the dipping sauce is a light addition. This is a perfect starter for a honey-themed, start-of-the-year dinner.

Here's a thought: Have an hors d'oeuvres party using all the hors d'oeuvres in the book. Fun!

The ingredients:

FOR THE CHICKEN

- 1 **Meyer lemon**
- 2 **garlic** cloves, crushed
- ½ teaspoon sea salt
- ½ teaspoon freshly ground black pepper
- 2 tablespoons extra-virgin olive oil
- 1 tablespoon **honey**, preferably orange blossom honey
- 1 tablespoon **red wine**
- 1 pound boneless, skinless chicken breasts, cut into 1-inch cubes

FOR THE DIPPING SAUCE

- 2 cups plain Greek yogurt
- 2 tablespoons **honey**, preferably orange blossom honey
- Zest of 1 **orange**
- 2 tablespoons **orange juice**

FOR THE SKEWERS

- 25 (8-inch) wooden skewers
- ¼ pound button mushrooms, stems removed
- ¼ pound **pearl onions**, peeled (you can make it easy by purchasing frozen peeled onions)
- 1 **red bell pepper**, seeded and cut into 1-inch squares

Here's what you do:

1. Juice the lemon. Combine the juice of the lemon and its rind (yes, put the entire lemon rind in the marinade), the garlic, salt, pepper, olive oil, honey, and red wine in a medium bowl. Whisk together. Add the chicken cubes and allow to sit, covered and refrigerated, for 1 hour. Stir occasionally.

2. For the dipping sauce, combine the yogurt, honey, orange zest, and orange juice in a small bowl, stirring until well mixed. Refrigerate until needed.

3. Soak the skewers in water for at least 30 minutes to prevent them from burning. Prepare a medium-hot fire in a charcoal or gas grill, or preheat the oven broiler. (I prefer grilling, as the char of the grill will enhance the overall flavor of the skewers.)

4. Skewer the mushrooms, pearl onions, bell pepper, and chicken, putting the chicken on last. Be sure to leave part of each skewer empty at the end for your guests to hold.

5. Grill for about 3 minutes per side, rotating as each side is cooked. Place a strip of aluminum foil under the empty portion of the skewers to ensure they do not burn. Or, broil for about 3 minutes per side.

6. Serve with the dipping sauce.

Serves 6 as an appetizer

a bee's lifespan

In the summer during high productivity, a worker bee lives for 42 to 45 days. In a cold winter, a worker could live for 6 months. The queen can live for 4 or 5 years. Some say she ceases to be productive after 2 years, but that is a matter of considerable controversy.

bold = *foods pollinated or produced by bees*

What is a Honey Varietal?

Here's the simple answer: Bees make honey by collecting nectar from flowers and combining that nectar with an enzyme that converts the sucrose in the nectar to glucose and fructose. The bees put the liquid, which at that point is 87 percent water, into empty honeycomb cells.

Worker bees fan the uncapped cells until the liquid evaporates and the resulting liquid is 17 percent water, the consistency we are accustomed to in honey. At that point, the bees put a wax cap on the cell, which keeps the honey pure until the cap is removed.

A specific honey varietal is the result of a beekeeper paying close attention to what is in bloom in the area around his or her beehives. Many times the beehives are positioned in the middle of a blooming crop, such as orange blossoms. The bees collect nectar from the orange blossoms, filling the empty honeycomb with orange blossom nectar. When the orange trees are finished blooming, the beekeeper collects the filled honey frames and takes them to a safe storage place until it is time to extract the honey. This ensures that no other nectar will be mixed in and that the honey will remain a single varietal: orange blossom honey. In the United States alone, there are more than 300 specific honey varietals.

Many beekeepers do not move their hives or try to collect single varietals, simply allowing their bees to forage in a two- to five-mile radius around their hives. In this case, the honey is called "spring honey" or "mixed wildflower honey." I call my honey "Stoney Knob Gold," after my street and in acknowledgment of the fact that my bees' honey is from my neighborhood. The miracle of a mixed honey is that it is a true reflection of the flowers and trees in bloom in the neighborhood of the beehives. The first time I tasted the honey made by my bees, I almost fainted with giddiness as I inhaled the aromas of my home flowers, rolling the flavors around in my mouth. My bees, true artists, had created a unique honey.

tuscan tomato soup with orange slices

I adore this soup. I often make it at home when I suddenly find that company is coming. It is delicious, easy, and unusual enough to make your guests praise you — always a good thing! In my shop, we frequently make this when we need something quick and easy for the day's soup. Give it a try. I'm sure you'll soon add it to your repertoire.

The ingredients:

- 4 **garlic** cloves, peeled
- 3 tablespoons extra-virgin olive oil
- 1 small **red onion**, cut into thin slices
- 1 **red bell pepper**, seeded and cut into large chunks
- 1 **yellow bell pepper**, seeded and cut into large chunks
- 2 (28-ounce) cans whole tomatoes (San Marzano are great)
- 1 **navel orange**, cut into wedges and then cut into thin slices (leave the rind on!)

    Coarse salt and freshly ground black pepper

- 1 cup shaved Parmesan cheese

Here's what you do:

1. Sauté the whole garlic cloves in the olive oil in a medium soup pot over low heat until soft, about 3 minutes. Add the onion and bell peppers. Continue to cook until they are soft, about 5 minutes longer.

2. Add the tomatoes. I like a lot of texture in my soup, so I usually coarsely cut the tomatoes with a knife and fork once they are in the pot. Add the orange slices. Simmer for 20 minutes or so.

3. Season with salt and pepper to taste. Some canned tomatoes are very salty and others less so. I have a salt grinder in my kitchen and give the soup a few grinds of salt and pepper before serving. The oranges will be soft and will add a surprising taste.

4. I love the chunkiness of this soup, but if you prefer a smoother texture, transfer some or all of the soup in batches to a food processor or a blender, pulsing until you arrive at your favorite texture. Or blend the whole mixture in the pot with an immersion blender.

5. Garnish with the shaved Parmesan and get ready for compliments!

Serves 8

papa's salad with clementines

My grandfather ran a newspaper store. My grandmother was the cook in the family and the person who inspired my mother to cook. Any time you see one of my recipes with "Mama's" in the title, it is one of hers. But every once in a while my grandfather, Papa, came up with something, and it became his. This salad is one of those recipes, though I have adapted it slightly, changing the sugar in the original to orange blossom honey. Give it a go. I'm certain you'll be pleased with its crisp sweetness.

The ingredients:

FOR THE SALAD

- 1 head Boston lettuce (or local Bibb)
- 1 head baby romaine lettuce
- 1 **cucumber**, peeled and sliced into thin rounds
- 1 small **sweet onion**, sliced into thin rounds
- 6 **radishes**, sliced into thin rounds
- 2 seedless **clementines**, peeled and sectioned, white parts removed
- 2 hard-boiled eggs, peeled and sliced into rounds

FOR THE DRESSING

- 2 tablespoons **honey**, preferably orange blossom honey
- 2 tablespoons white vinegar
- ½ cup half-and-half
- Coarse salt and freshly ground black pepper (optional)

Serves 8

Here's what you do:

1. Tear the Boston and romaine lettuce into large pieces.

2. Combine the lettuces, cucumber, onion, radishes, clementines, and eggs in a large decorative ceramic or wooden bowl.

3. To make the dressing, combine the honey and vinegar in a small bowl, stirring until thoroughly blended. Add the half-and-half, whisking until all is well combined.

4. Drizzle the salad dressing over the salad just before serving (and not before), tossing very gently. Papa did not add salt or pepper, but if you wish, sprinkle some coarse salt and grind some fresh pepper over the top right at the end.

5. Take the bowl to the table and tell your guests about my grandfather if you like. Enjoy!

Try in Summer

This is a fine early winter salad because all of these ingredients can be found then, but if you have a good farmers' market, you might want to try this salad in the middle of your local growing season for a really fresh treat.

bold = *foods pollinated or produced by bees*

pork tenderloin with orange blossom honey mustard

Pork is well complemented by sweet things. On a trip to Tuscany, I was wowed by a dinner of local pork chops served with a sweet onion confit made with sugar. The flavor stayed with me and, after making my version of that dish a number of times, I decided to play with it, using honey and fresh fruit. Here's what I came up with.

The ingredients:

- 2 pork tenderloins (about 2 pounds total)
- 2 tablespoons extra-virgin olive oil
- ½ teaspoon granulated **garlic**
- ½ teaspoon kosher salt
- ½ teaspoon freshly ground black pepper
- 2 tablespoons **Dijon mustard**
- 2 tablespoons **honey**, preferably orange blossom honey
- 1 **navel orange**
- Juice of 1 **Meyer lemon**

Here's what you do:

1. Preheat the oven to 375°F.

2. Drizzle the pork with 1 tablespoon of the oil. Season with the garlic, salt, and pepper.

3. Drizzle the remaining 1 tablespoon of oil into a cast-iron skillet over medium-high heat, swirling to coat the surface of the pan. When the oil is hot, add the tenderloins. Turn when browned, after about 1 minute, and cook for 1 minute longer. Ideally, each tenderloin will curve close to the sides of the pan, leaving a space in the center.

4. Remove the skillet from the heat.

5. Combine the mustard and honey in a small bowl and blend well with a fork. Spoon the mixture over the pork.

6. Cut the orange into thin slices, leaving the rind on. Layer them in the skillet, lining them up in the center of the pan, overlapping to cover all the spaces in the pan.

7. Roast the pork in the oven for 15 to 20 minutes, or until the internal temperature is 145°F. Remove from the oven and squeeze the lemon juice over the pork. Let stand for at least 5 minutes before slicing.

8. Slice the tenderloins in diagonal slices, about 1 inch thick. Place a slice or two of the cooked orange on the plate and put the sliced pork on top. Spoon the pan drippings over the sliced pork and serve. Oh my!

Serves 6

bold = *foods pollinated or produced by bees*

oven-roasted brussels sprouts with garlic

Even though my mother was a wonderful cook, she did not usually stray from her well-beaten path. Vegetables in our house were steamed. And in fact, most vegetables started in the frozen state. Steaming frozen vegetables is okay if that's all you have on hand, but things have changed dramatically since my mother's time, with fresh, local food much more available. Oven roasting is such a fine way to bring out the flavors of vegetables that I rarely steam anything anymore. There is no honey in this recipe, but without bees, we wouldn't have Brussels sprouts (or garlic).

If you are lucky enough to live in a place with local farmers' markets, you'll be thrilled to buy a stalk of Brussels sprouts, snip them off the stalk, and roast them right away. Second best is to buy a bag full of bright green sprouts in the produce section of your grocery store. Roasting them will bring out their sweetness.

The ingredients:

- 1 stalk **Brussels sprouts** (about 1 pound)
- 4 **garlic** cloves, chopped into large pieces
- ½ cup extra-virgin olive oil
- 1 teaspoon kosher salt or coarse sea salt

Here's what you do:

1. Preheat the oven to 375°F.

2. Cut the Brussels sprouts off the stalk, trimming off any bruised outer leaves. Cut each sprout in half lengthwise. Spread the cut sprouts on a sturdy baking sheet.

3. Spread the chopped garlic over the sprouts. Drizzle with the olive oil and sprinkle with the salt.

4. Bake for 20 to 25 minutes, keeping watch toward the end of the baking time, until slightly browned and tender when poked with a fork. Taste and add another bit of salt if desired.

5. Serve immediately.

Serves 4–5

citrus smoothies

Smoothies make a great start to the day. To make one, you simply need fresh fruit, yogurt, honey, and a blender. During our recipe-testing sessions, we realized that this colder version, with ice, was significantly better than the version we made without ice. I have one of those super-duper blenders that cranks through ice cubes as if they were butter. If you don't, you may need to chop up whole ice cubes before adding them to your blender (put them in a heavy-duty plastic bag and pound them gently with a hammer).

The honey adds its own intrigue to the taste.

The ingredients:

- 1 medium banana
- ½ cup **strawberries**
- Zest from 1 **orange**
- 1½ cups plain Greek yogurt
- 1 cup **orange juice**
- ¼ cup **honey**, preferably orange blossom honey
- 1 teaspoon vanilla extract
- 1½ cups ice cubes
- 4–6 whole **strawberries** for garnish

Here's what you do:

1. Combine the banana, strawberries, orange zest, yogurt, orange juice, honey, vanilla, and ice in a blender and pulse until thoroughly mixed. Easy enough!

2. Pour into glasses and garnish each with a fat red strawberry. Serve immediately.

Serves 4–6

bold = *foods pollinated or produced by bees*

coconut macaroons with dried cherries

I always thought macaroons were a big deal. But these, my friends, are easy. Play around with the dried fruit, using your favorite. The honey and butter drizzle adds a nice crispy edge to the soft centers, and the tiny bit of salt balances the sweetness. If you want to be really fancy, melt some chocolate and dip each one halfway into it. Zounds!

The ingredients:

- 1 cup unsweetened flaked **coconut**
- 1 cup sweetened flaked **coconut**
- 8 egg whites
- Salt
- ¼ cup dried **cherries**
- 2 tablespoons butter
- 1 tablespoon **honey**, preferably orange blossom honey

Here's what you do:

1. Preheat the oven to 350°F.

2. Combine the unsweetened and sweetened coconut on a baking sheet. Lightly toast in the oven for 5 to 10 minutes. Keep close watch so the mixture does not burn, though you do want a nice toasted golden brown color. Remove from the oven and set aside to cool.

3. Reduce the oven temperature to 325°F. (If using a convection oven, leave at 350°F.)

4. Whip the egg whites with a pinch of salt in a medium bowl until the whites stiffen into firm peaks.

5. Fold the toasted coconut into the egg white mixture.

6. Line a baking sheet with parchment paper. Drop tablespoon-size rounds (I prefer to use a small ice cream scoop) of the coconut mixture onto the baking sheet. Press one or two dried cherries into the top of each macaroon.

7. Melt the butter and honey together in a microwave on high for 20 seconds. Drizzle the mixture over the top of each macaroon. Sprinkle with a tiny pinch of salt.

8. Bake the macaroons for 20 to 25 minutes, until lightly browned (watch carefully to avoid burning!), or for 8 to 10 minutes if using a convection oven. The macaroons should be dry to the touch. You may need to bake them longer if it is a humid day. If you live in a dry area, these will keep well for a week — if the cookie patrol doesn't find them first.

Makes 25–30 small cookies

bold = *foods pollinated or produced by bees*

tupelo

The pedigree of honey
Does not concern the bee;
A clover, any time, to him
Is aristocracy. — EMILY DICKINSON

Tupelo honey is a unique honey that comes from a very specific region of Florida and Georgia. The Tupelo gum tree blooms only during April and May, making the honey a prized rarity. Beehives are placed along the sides of the Ogeechee, Apalachicola, and Chattahoochee Rivers during those two months. The bees collect nectar from the white Ogeechee tupelo tree, turning it into this light, delicious, buttery honey. Interestingly, due to its unique sugar chemistry, tupelo honey is one of the few honeys that does not crystallize.

COLOR
Medium light amber

SMELL
Stewed fruit; raisins, apricots, and prunes mixed with floral scents

TASTE
Gently caramelized; lightly sweet, buttery

AFTERTASTE
None

leaneau's grilled pineapple skewers

My friend Lea brought these to a Thanksgiving party at my house a few years ago. That year I invited all my friends to bring something representative of a Thanksgiving they had spent in an unusual location. Lea lived on a small sailboat for a few of her younger years and had been in the Caribbean during Thanksgiving one year. This was her offering at our dinner, and it has been a part of my party thoughts ever since. I've adapted the original recipe by making it with tupelo honey.

The ingredients:

- 2 **garlic** cloves, chopped
- 1 teaspoon minced fresh gingerroot
- ½ cup soy sauce
- ¼ cup **honey**, preferably tupelo honey
- 2 tablespoons **lime juice**
- 2 tablespoons **white wine** or **sweet sherry**
- 1 tablespoon toasted **sesame oil**
- 1 whole pineapple, peeled, cored, and cut into 1-inch chunks
- 24 (8-inch) bamboo skewers

Here's what you do:

1. Combine the garlic, gingerroot, soy sauce, honey, lime juice, white wine, and sesame oil in a medium, nonreactive bowl. Whisk well to thoroughly combine. Add the pineapple chunks. Stir well, coating all the chunks with the marinade. Allow to sit for 1 hour, either refrigerated or at room temperature.

2. Soak the skewers in water for at least 30 minutes to prevent them from burning. Prepare a medium fire in a charcoal or gas grill, or preheat the broiler.

3. Skewer the marinated pineapple, using 2 or 3 chunks per skewer. Grill, turning as needed, until all four sides of the pineapple are browned; or broil for about 2 minutes per side.

4. Serve immediately, passing to your guests, or, in the case of an outdoor grill, letting them take the skewers right off the grill if you like.

Makes 24 skewers

bold = *foods pollinated or produced by bees*

mama's winter vegetable soup

Though she died before I was born, my grandmother's cooking lived on in my early life and, truthfully, still does. Of course my mother adopted and adapted many of her mother's recipes; now they have entered my repertoire.

Here's a wonderful vegetable soup that works no matter what time of year you are cooking. The most important part is, of course, to use local, fresh vegetables. And in this case, let's concentrate on vegetables that need bees for pollination. Do feel free to alter the vegetables according to what is the freshest when you go to the market (you can check the list of bee-dependent vegetables page 195). This soup improves with age, so you may make it a day before you want to serve it.

The ingredients:

- 3 medium **sweet onions**, peeled and cut into thin round slices
- 3 **carrots**, peeled and cut into diagonal slices
- 1 **turnip**, peeled and cut into ½-inch cubes
- 1 **leek**, well washed and cut into 1-inch pieces (do not use the toughest tops of the green part)
- 1 cup chopped **celery**, including tops if possible
- 4 fresh curly **parsley** sprigs
- 1 teaspoon kosher or sea salt
- ½ cup **haricots verts** (baby French green beans) or **green beans**, cut into ½-inch pieces
- ½ cup **broccoli** pieces, separated into ¼ cup florets and ¼ cup sliced, peeled stems
- ½ cup **peas**
- ½ **red bell pepper**, seeded and cut in slivers
- 1 tablespoon butter
- ½ cup shaved or grated Parmesan or Grana Padano cheese

Here's what you do:

1. Combine the onions, carrots, turnip, leek, celery, and parsley in a large soup pot. Cover with about 2 quarts water and bring to a boil. Add the salt. Reduce the heat to a simmer. Cook until the vegetables are tender, 30 to 40 minutes.

2. Add the haricots verts and broccoli stems and cook for 15 minutes.

3. Add the broccoli florets, peas, and bell pepper, and cook until they are tender, about 15 minutes. Stir in the butter.

4. Ladle the soup into bowls and serve with the shaved or grated Parmesan on top.

Serves 6–8

bold = *foods pollinated or produced by bees*

What's Inside Those Beehives?

Man-made beehives are wonderfully simple and also amazingly complex structures. They are specifically designed to mimic certain qualities of wild beehives, and also to make it as easy as possible for beekeepers to care for the bees and harvest their honey.

You have probably seen the stacked boxes in a field or even on a rooftop or in a backyard. At the bottom of the stack is the deepest box, known as the "hive body" or the "nursery box." This is where the colony of the hive is raised.

Stacked on top of the hive body are the "supers," the boxes where the honey is made and stored. Supers are available in three different depths: deep, medium, and shallow.

The hive body and the supers accommodate 10 moveable, empty wooden frames. The bees fill the frames with honeycomb to use for raising brood (baby bees) or for storing honey and pollen. Shallow supers have shallow frames, medium supers have medium frames, and — yup — deeps hold deep frames.

A beekeeper chooses the size of the super based on a number of factors, with the most important being his or her physical strength. A shallow super filled with honey weighs 35 pounds, a medium 50 pounds, and a deep up to 85 pounds. As you might imagine, someone like me (not so big and not so strong) is probably not going to haul around 85-pound boxes! Bigger and stronger beekeepers use medium or deep supers, but most hobby beekeepers use shallow supers.

Langstroth hives (the most popular type) respect what is called "bee space": Bees like to have ⅜ inch of personal space between honeycombs. When the frames inside a Langstroth hive are filled with honeycombs, the space between the honeycombs

in each frame is ⅜ inch, and there is ⅜ inch between the outside edge of the frames and the inside of the boxes. If you ever get the chance to see a picture of a wild honeybee colony, you will see fans of honeycomb with a tidy ⅜ inch between each fan. Our Langstroth hives merely attempt to mimic what the bees do naturally.

THE PARTS OF THE BEEHIVE

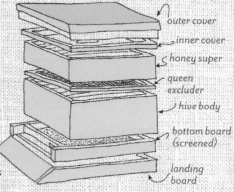

outer cover

inner cover

honey super

queen excluder

hive body

bottom board (screened)

landing board

creamy chicken and coconut curry

This is one of my mother's recipes that I've adapted to include honey (of course!) and to accommodate our lighter styles of eating. It is a fine dish for a dinner party, especially one that includes serving a main course from a chafing dish on a buffet. The flavors improve over time, so feel free to make it ahead of time, freeze it, and defrost and serve when you have no energy to cook. You could use premade broth, but this version is really not that hard and is so good that I recommend you go the extra step to start from scratch. This mild curry goes nicely with rice.

The ingredients:

FOR THE CHICKEN AND STOCK

2 **carrots**

1 medium yellow **onion**, peeled

1 cup chopped **celery**, with tops

1 bay leaf

4 whole peppercorns

2 tablespoons kosher salt

1 whole chicken (about 5 pounds), cut into parts

FOR THE COCONUT CURRY

4 tablespoons unsalted butter

6 tablespoons unbleached all-purpose flour

2 **garlic** cloves

1½ teaspoons curry powder

½ teaspoon paprika

1 tablespoon **honey**, preferably tupelo honey

½ cup half-and-half

½ cup unsweetened flaked or shredded **coconut**, plus more for garnish (optional)

Here's what you do:

1. Add 3 quarts water to a large pot. Add the carrots, onion, celery, bay leaf, peppercorns, and salt. Bring to a boil, then immediately reduce the heat to a simmer. Add the chicken parts.

2. Cook until the chicken is quite tender, 1 hour or more. Remove from heat and let cool. (**NOTE**: A quick way to cool hot chicken stock is to put the hot pot in a sink filled with ice water. When the sink water warms up, replace it with another round of ice water; repeat until the stock is cool.)

3. Pull the chicken off the bones, removing and discarding the skin. Return the chicken bones to the stock and simmer for another 30 minutes or so. Set the chicken meat aside, refrigerating it if you are not going to complete the recipe immediately.

4. Strain the stock into a bowl and set it aside too, refrigerating it if you are not going to complete the recipe immediately.

5. To make the coconut curry, melt the butter over medium heat in a large pot. Turn the heat to low when the butter has melted.

6. Add the flour and blend with a wooden spoon, stirring until it is a smooth, thick paste. Add the garlic, curry powder, paprika, and honey, and stir until thoroughly combined.

7. Slowly add the half-and-half and then 2 to 3 cups of the stock, gently stirring until the sauce is smooth and nicely thick, about the consistency of *honey*!

8. Fold in the reserved chicken and the coconut just before serving, and warm thoroughly. Add more stock if the consistency is too thick for you.

9. Serve with additional coconut sprinkled on top if you wish.

Serves 8

Measuring Honey

When measuring honey, you may want to first dip your measuring spoon or cup in a light vegetable oil or quickly spray it with cooking spray. The honey will then easily drip off the spoon or cup.

sweet and salty kale crisps

What a fine recipe this is! It's easy. It's delicious. And you'll want to make it over and over again. I have been eating oven-roasted kale for some time, prompted by a farmer friend who suggested that I spritz the kale leaves with olive oil and bake them in a slow oven until they crisp up. I recently wondered what would happen if I added a little sweetness to the mix. What a fine discovery that was. I got sweet, sour, salty, and bitter tastes all in one bite — gustatory perfection. Try this and see what I mean.

Serves 2–3 as a side or 5–6 as a cocktail-hour nibble

The ingredients:

- 1 bunch curly **kale** (you can use other kinds of kale too; have fun experimenting)
- 2–3 tablespoons extra-virgin olive oil
- ⅛ teaspoon kosher salt
- ⅛ teaspoon freshly ground black pepper
- 1 tablespoon **honey**, preferably tupelo honey
- 1 tablespoon **red wine** vinegar

Here's what you do:

1. Preheat the oven to 350°F.

2. Tear the kale leaves away from the stems. Put the stems in your compost bucket. Tear any large leaves into smaller pieces, 2 to 3 inches each.

3. Arrange the torn leaves on a baking sheet and spray with olive oil. (I keep olive oil in a spray bottle, the kind you might use to spritz plants.) If you don't have a spray bottle, toss the kale and oil together in a bowl before spreading the leaves on the baking sheet. Season with the salt and pepper.

4. Bake until crisp, about 20 minutes, tossing once or twice. Remove from the oven, drizzle the honey over the kale, and bake for 5 minutes longer. Remove from the oven and sprinkle the leaves with the red wine vinegar.

5. Serve immediately, letting guests help themselves to this crispy finger food.

bold = *foods pollinated or produced by bees*

israeli couscous with fresh fruit

Israeli couscous is fun to cook and eat. Different from traditional couscous, the finished grains of which are tiny, Israeli couscous cooks up into small, smooth, round beads. We often serve this as a room-temperature salad in my café, but the salad can be served warm too. Another big plus is that it cooks very quickly and is a great starch. This is a simple recipe, but feel free to add freshly roasted seasonal vegetables and fresh herbs from your garden or your windowsill (if you're making this in the dark of winter).

The ingredients:

Pinch of salt

1   cup Israeli couscous (also called pearl couscous)

¼   cup sliced tamari **almonds**

¼   cup **raisins**

1   **orange**, peeled and cut into sections

1   **scallion**, cut into thin slices

¼   cup extra-virgin olive oil

2   tablespoons **apple cider vinegar**

Serves 8

Here's what you do:

1. Bring 1¼ cups of lightly salted water to a boil in a small pot. Add the couscous, stir, cover the pot, and reduce the heat to a simmer. Cook, stirring occasionally, until the water is absorbed, 5 to 8 minutes. Remove the pot from the heat and allow to cool.

2. Combine the couscous, almonds, raisins, orange sections, and scallion in a 1-quart bowl. Drizzle with the olive oil and vinegar and toss. Enjoy!

baked bananas with honeyed cream

I grew up with baked bananas. Ours was not a big dessert-eating family, but this dish made it to our table often, especially on dark winter nights. It makes me warm just thinking about it. My mother used brown sugar as the sweetener in the creamy sauce, but now I like to use honey. Tupelo honey blends nicely with the cream and is a not-too-sweet addition to the bananas.

The ingredients:

- 4 ripe bananas
- 4 tablespoons salted butter, cut into pieces
- 2 tablespoons light brown sugar
- 1 cup **orange juice**
- 1 cup sour cream
- 1 tablespoon **honey**, preferably tupelo honey
- 1 teaspoon vanilla extract

Here's what you do:

1. Preheat the oven to 375°F.

2. Peel the bananas and cut them in half lengthwise. Place the bananas in a baking dish that is nice enough to bring to the table. Dot with the butter and sprinkle with the brown sugar.

3. Carefully pour the orange juice into the bottom of the baking dish, trying not to dislodge the butter and sugar.

4. Bake the bananas for 15 minutes, or until they are lightly browned and the liquid in the bottom of the dish is bubbly. Remove from the oven and allow to cool slightly.

5. Combine the sour cream, honey, and vanilla in a small bowl.

6. Place the baking dish on the table and arrange the bowl with the sour cream sauce on the side. Invite guests to serve themselves, each taking half a banana, some juice, and a dollop of the sauce.

Serves 8

---

bold = *foods pollinated or produced by bees*

mango–key lime slushes

Here's a very refreshing drink. The key lime juice is so much more interesting than regular lime juice, adding a subtle and curious background taste. If you can't find fresh key limes, you can probably find the juice, already squeezed, in your grocery store near the cocktail mixing area (near the cherries and olives and fancy stirrers). For best results, make sure the mangos are very ripe and sweet.

For the mint garnish, I suggest pineapple mint for something quite lovely and complementary to the flavors in this drink, though a chocolate mint could be fun too. If you can't find these varieties, regular mint works fine.

The ingredients:

   3 cups chopped **mango**

   1 cup **orange juice**

   ⅓ cup **key lime juice**

   ¼ cup **honey**, preferably tupelo honey

   ⅛ teaspoon salt

4–5 cups ice cubes

   Fresh **mint** sprigs for garnish

Here's what you do:

1. Combine the mango, orange juice, key lime juice, honey, and salt in a blender. Pulse until well blended and thoroughly puréed.

2. Add the ice cubes and continue to pulse until the mixture is an icy blend. This will take 10 to 15 seconds with a good, strong blender. A Vitamix mixer will do the trick in about 5 seconds.

3. Pour into glasses, and garnish with a sprig of fresh mint. Serve immediately.

Serves 6–8

bold = *foods pollinated or produced by bees*

acacia

The careful insect midst his works I view,
Now from the flowers exhaust the fragrant dew;
With golden treasures load his little thighs,
And steer his distant journey through the skies.

— JOHN GAY, "RURAL SPORTS"

Acacia honey is made from flowers of the false acacia, also known in the United States as the black locust tree. This tree, native to the northeastern United States, is also widely known in Europe.

Acacia honey is available here in specialty stores. It is mainly produced in Bulgaria, Hungary, and Romania, but it can easily be found in France and Italy too. When you are traveling or are at a specialty honey store, do pick some up! It will be the lightest honey in your collection and will be a lovely addition to your morning tea, to toast, or to any of the recipes in this chapter. And, incidentally, acacia honey is said to be one of the best wound-healing honeys.

COLOR	SMELL	TASTE	AFTERTASTE
Very light yellow	Very mild	Subtle, fresh, normally sweet; buttery vanilla	Very mild and pleasant caramel

roasted baby vegetables with a dilled cream dipping sauce

Early spring is when you can start looking for the newest of baby vegetables in the markets. We're a bit too far north in Asheville to get produce from our own gardens, but we do get regional vegetables in our stores and, inspired, I like to roast them with some coarse salt and serve them to my friends with a nice herbed dip. There's no honey in this recipe, but honeybees pollinate *all* of the vegetables I've listed. Thank you, honeybees.

Tell your guests the "every third bite" story behind your assortment (see page 10), and your presentation will lift itself above the usual vegetable tray. Refer to the list on page 195 for other vegetables that you can include if you make this in a different season.

The ingredients:

FOR THE VEGETABLES

- 1 bunch (about 1 pound) baby **carrots**, whole if possible, tops removed
- 1 bunch (about 1 pound) baby **beets**, tops removed, but leave 1 inch of the stem
- 1 bunch (about 1 pound) baby **turnips**, tops removed, but leave 1 inch of the stem
- ¼ cup extra-virgin olive oil
- 1 bunch (about ½ pound) **radishes**, tops removed, but leave 1 inch of the stem
- ½ pound spring **peas** (optional)

FOR THE DIPPING SAUCE

- 2 cups plain Greek yogurt
- 1 tablespoon fresh **dill**
- Zest of 1 **lime**
- Juice of 1 **lime**
- Sea salt

Here's what you do:

1. Preheat the oven to 375°F.

2. Cut any carrots that are more than 4 inches long into 2-inch-long diagonal slices.

3. Cut any large beets and turnips into bite-size pieces.

recipe continues on following page >

4. Toss the carrots, beets, and turnips separately in the olive oil and spread out on a large baking sheet, keeping the vegetables separated from each other. Bake for 20 to 30 minutes (tiny vegetables will take less time), checking often toward the end to make sure they don't burn. The vegetables should be tender but not overly soft. Allow to cool.

5. Cut any large radishes into bite-size pieces.

6. If using peas that are not really fresh, bring a small pot of water to a boil and immerse them in the boiling water until they turn bright green, about 2 minutes. Remove and immediately plunge in a bowl filled with ice water to stop the cooking and keep them bright green. If peas are really fresh, leave uncooked.

7. To make the dip, combine the yogurt, dill, lime zest, and lime juice in a medium bowl. Add a pinch or two of sea salt.

8. Present the vegetables on a serving platter. I like to make little mounds of each item and arrange them in order of the color spectrum (red, orange, yellow, green, blue, indigo, and violet), but a random scatter is fun too. I like to garnish with something that appears in the recipe, so I sometimes use the tops from the vegetables as a base or to tuck around the bowl holding the dip.

*Serves 8
as an appetizer*

**sweet
fact**

It takes 500,000 flakes of wax to make 1 pound of beeswax.

elsie's rhubarb soup

After my mother wrote *The Blueberry Hill Cookbook* and Blueberry Hill Inn became more popular, readers and visitors to the inn sent her their favorite recipes. Many of them became our favorites. Here's one such recipe. Now that it has been modified to incorporate honey rather than sugar, it is one of my favorites, though it is unusual. You might want to consider it a dessert soup. As soon as your rhubarb pokes up enough to cut or as soon as it appears in your markets, buy some and make this soup.

NOTE: The egg yolk in this recipe thickens the soup. If you are nervous about using a raw egg, leave it out, but I feel confident using a fresh, local, organic egg.

The ingredients:

- 3 cups thinly sliced rhubarb (in a pinch, it is okay to use frozen as long as it is plain frozen rhubarb)
- 1 teaspoon cornstarch
- ½ cup **honey**, preferably acacia honey or the lightest local honey you can find
- ½ cup chilled whipping cream
- 1 organic egg yolk

  Zest of 1 **lemon** for garnish

Here's what you do:

1. Cover the rhubarb with cold water in a medium, nonreactive saucepan and bring to a boil. Reduce the heat to a simmer and cook until the fruit is very soft, about 15 minutes or more.

2. Press the rhubarb through a sieve or, better, a food mill. This will keep any tough, stringy parts of the rhubarb out of the soup. Toss the strings in the compost. Return the rhubarb mixture to the saucepan.

3. Combine the cornstarch with ¼ cup cold water in a small cup and mix well. Add the cornstarch to the rhubarb mixture, stirring until there is no cloudiness (from the cornstarch mixture).

4. Add the honey to the cornstarch and rhubarb mixture. If the rhubarb is very tart, you may like to add a bit more honey, but don't overdo it. This soup is memorable for not being too sweet. Chill thoroughly.

5. Whip the chilled cream and the egg yolk together in a medium bowl until stiff peaks form (be careful — you do not want to make butter!). Add to the soup and combine well.

6. Garnish with the lemon zest, and serve.

Serves 8–10

mushroom salad with fresh dill

I love this recipe! It's perfect for an early spring luncheon or as an accompaniment for a spring dinner. It works best if made a few hours before serving. The original version was made with sugar. I've replaced that with March's varietal, acacia honey, which adds a sweet depth and complexity that is so much more interesting. Do give this one a try.

The ingredients:

- 1 pound button mushrooms
- 1 tablespoon chopped fresh curly **parsley**, plus more for garnish
- 1 teaspoon sea salt
- ½ teaspoon **celery seeds**
- ½ teaspoon fresh **dill**, finely chopped
- ⅛ teaspoon granulated **garlic**
- 1 cup heavy cream
- Zest of 1 **lemon**
- Juice of 1 **lemon** (about 2 tablespoons)
- ½ teaspoon **honey**, preferably acacia honey
- 1 head Bibb lettuce

Serves 4–5

Here's what you do:

1. Trim off the bottoms of the mushroom stems if they are dirty. Cut the mushrooms into thin "umbrella" slices.

2. Combine the mushrooms, parsley, salt, celery seeds, dill, garlic, heavy cream, lemon zest, lemon juice, and honey in a 1-quart bowl and stir until the mushrooms are completely coated with the cream mixture.

3. Refrigerate for at least 1 hour, if possible.

4. Arrange a few lettuce leaves on each salad plate and place 1 cup of the mushrooms in the lettuce. Garnish with extra parsley, and serve.

bold = *foods pollinated or produced by bees*

the Queen

There are three different castes of bees in a bee colony: the queen, the drones, and the workers. The queen is the head of the colony. Bees can live without a queen, but the hive cannot go on for very long without her, nor can it grow and thrive.

The queen's primary job is to lay eggs, which she does all day long, every day, during the warm times of the year. Amazingly, she can lay up to 1,500 eggs each day, which, when totaled, weigh more than she does. She is the only bee who can lay eggs that will hatch into functional members of the colony.

She is the largest bee in the colony and is fairly easy to spot. She is usually surrounded by a coterie of workers who groom her, feed her, and take care of her. The queen has a unique scent, called a pheromone, which is spread throughout the colony by those workers, making all the bees know that she is their leader and this is their home.

The queen has an effective life span of up to five years, though some beekeepers think that she stops being productive after two years. Those beekeepers replace her, either by breeding a new one or by purchasing a new queen from a breeder. Worker bees will also "make" a new queen once they realize that their queen has gone or is not doing a good job.

To "make" a new queen, worker bees will create a new, large cell for a few eggs. Other workers, seeing the special, larger queen cell, will feed those eggs a diet of "royal jelly," a rich food that all bee babies get for a short time, but that the queen gets for many more days. Sadly, it is a survival of the fittest contest here: The first queen to hatch will take it upon herself to destroy the others. If, by chance, more than one hatches, the strongest one wins and takes over the hive.

A newborn queen spends her first few days inside the hive and then goes on "mating flights" — brief forays to an area in the air in the vicinity of the hive. There, in what are known as "drone congregation areas" (I think of them as bee singles bars), the queen connects and mates with a number of drones from an assortment of hives. Over a period of a day or two, the queen mates with an average of 12 different drones, taking on the genetic material of each. When she returns to her home hive after these mating flights, she stays inside the hive for the rest of her life, laying eggs that are genetically diverse.

whole roasted chicken with fresh herbs

My sister gave me a ceramic chicken baker that is a fancified substitute for baking a chicken over a beer can. That beer-can kind of chicken is great, though the paint from the cans is not so great. But I digress. This chicken recipe works well at Passover, an Easter dinner, a spring dinner, or any other dinner you might have — and you don't need a special baker. Here's my guide for you.

The ingredients:

- 1 farm-fresh chicken (3–4 pounds)
- 2 **lemons**, cut in half
- 2 sweet medium **onions**, cut into large wedges
- 6 **garlic** cloves, peeled
- 4 fresh rosemary sprigs
- 4 fresh thyme sprigs
- 1 pound baby potatoes
- 2 **carrots**, peeled and cut into 1-inch diagonal slices
- 2 stalks **celery**, cut into 1-inch diagonal slices
- 1 teaspoon sea salt
- 1 teaspoon freshly ground black pepper
- 1 teaspoon granulated **garlic**
- 2 tablespoons unsalted butter, cut into pieces
- 2 tablespoons **honey**, preferably acacia honey

Here's what you do:

1. Preheat the oven to 375°F.

2. Wash the chicken and remove the giblets (you can use them for stock later, if you like). Dry the chicken and place it in a deep roasting pan (a 13-inch pan is good).

3. Tuck the halves of 1 lemon, the wedges of 1 onion, 2 of the garlic cloves, and 2 sprigs each of the rosemary and thyme into the cavity of the bird. In the bottom of the roaster, scatter the remaining onion, the remaining 4 garlic cloves, the potatoes, the carrots, and the celery. If the potatoes are larger than bite-size, cut them in half.

4. Season the chicken with the salt, pepper, and granulated garlic. Dot the bird with the butter.

5. Bake for about 1½ hours, or until the juices run clear when a small knife is inserted where the leg meets the body. If your chicken is larger or your oven is slower than mine, you may need to adjust the time, continuing to bake a bit longer.

6. Let the chicken rest for at least 15 minutes before cutting. This will allow the juices to stay in the meat instead of running out in the pan. While the chicken is resting, drizzle it with the honey.

7. Place the cooked vegetables on a platter and arrange the chicken, either whole or cut up, on top of the vegetables. Garnish with the remaining 2 sprigs each of rosemary and thyme. Squeeze the halves of the remaining lemon over the platter, pour any pan juices on top, and serve.

Serves 4–6

bold = *foods pollinated or produced by bees*

sparkling citrus punch

We make this punch all the time for parties. It is easy, refreshing, and quite adaptable. I encourage you to play with it to suit the season and your tastes. If you don't have freshly squeezed juice, reconstituted unsweetened frozen juice works just fine. If you use frozen concentrates, mix with two parts of water rather than the recommended three parts.

If not using an ice mold, simply fill the entire punch bowl with ice and garnish with the fresh fruit slices. The fresh fruit slices make this punch especially attractive.

The ingredients:

- 1 **lemon**, sliced into rings
- 1 **lime**, sliced into rings
- 1 **orange**, sliced into rings
- 20 whole **strawberries** (optional)
  Ice cubes as needed
- 1 (2-liter) bottle ginger ale
- 2 cups **lemon juice**
- 2 quarts **orange juice**
- 1 cup **lime juice**
- ¼ cup whole fresh **mint** leaves

Here's what you do:

1. Combine the lemon, lime, and orange slices and the strawberries, if using, in a ring mold the day before the party, completely filling the bottom of the mold. Cover with cold water and add more fruit to fill the mold. Do this in your sink, as excess water might spill over when you fill the mold with the fruit. Carefully transfer to the freezer (placing it on a baking sheet is helpful) and allow to freeze completely.

2. Just before your guests arrive, fill a large punch bowl half full with ice cubes. Combine the ginger ale, lemon juice, orange juice, and lime juice in the punch bowl.

3. To dislodge the ice ring, briefly run the outside of the mold under hot water, which will encourage the ice to slip out of the mold. Place the mold in the punch bowl. As it floats in the punch it will slowly melt, allowing the fruit to float in the punch. So pretty!

4. Garnish with the mint leaves.

Serves about 20

bold = *foods pollinated or produced by bees*

tzimmes

In the world of Jewish cookery there are numerous versions of this traditional dish. And, since it is a dish that can take a long time and can involve a lot of washing, peeling, chopping, and such, the word *tzimmes* has become synonymous with "a big deal."

Though I am technically Jewish (my mother was, which means I am), we did not grow up in a traditionally Jewish household or with this dish. Nevertheless, I have become fond of it as an accompaniment to a soothing, comfort-filled dinner of roasted chicken. It is traditionally a Passover dish, but really is a great part of any sort of dinner on a chilly night. Don't be intimidated by the "big deal." This one is not complicated to make.

The ingredients:

- 1 pound **carrots**, peeled and cut into 1-inch diagonal slices
- 1 pound sweet potatoes, peeled and cut into 1-inch cubes
- ½ pound Yukon Gold potatoes, cut into 1-inch cubes
- ½ cup dried **apricots**, cut in half
- ½ cup pitted **prunes**, cut in half
- ⅛ teaspoon ground **cinnamon**
- Zest of 1 **lemon**
- Juice of 1 **lemon**
- 2 tablespoons butter, cut into pieces
- ¼ cup **honey**, preferably acacia honey, plus more for drizzling
- ¼ cup slivered **almonds**
- 1 teaspoon kosher or sea salt
- Freshly ground black pepper

Serves 4

Here's what you do:

1. Preheat the oven to 350°F.

2. Combine the carrots, sweet potatoes, and Yukon Gold potatoes in a large soup pot and cover with cold water. Simmer until the vegetables are beginning to get tender, about 15 minutes.

3. Transfer the vegetables to a baking dish using a slotted spoon and add the apricots and prunes, scattering them around the vegetables. Sprinkle with the cinnamon, lemon zest, and lemon juice. Dot the mixture with the butter and drizzle the honey over all.

4. Cover with aluminum foil and bake for 30 minutes. Remove the foil, scatter the almonds on top of the mixture, and continue to bake for 10 minutes longer, or until a fork slides easily into the vegetables. Season with the salt, pepper to taste, and an additional drizzle of honey right before serving.

panna cotta with candied kumquats

To me, the most comforting of comfort foods is anything that has custard or cream or custard cream in it. *Panna cotta* means "cooked cream" in Italian, putting it at the top of my list of must-haves in the comfort department. Feel free to adapt the basic recipe. I've had a lovely lemony version served floating on a light cucumber soup, a cheesy version served with a midsummer corn chowder, and a version infused with Cointreau served as a very light finish to a winter feast. Inspired by a new crop of kumquats, I developed this recipe.

The ingredients:

FOR THE PANNA COTTA

1 cup whole milk

1 tablespoon unflavored powdered gelatin

3 cups whipping cream

⅓ cup **honey**, preferably acacia honey

Pinch of salt

FOR THE CANDIED KUMQUATS

½ cup **honey**, preferably acacia honey

4 cups sliced and seeded **kumquats** (sliced in half)

Here's what you do:

1. To make the panna cotta, pour the milk into a small bowl and sprinkle the gelatin on it, stirring just until blended. Allow to stand for 5 minutes to soften the gelatin.

2. Pour the mixture into a small heavy saucepan and warm over medium heat, allowing the gelatin to dissolve, about 5 minutes. Be careful not to let the milk boil.

3. Add the cream, honey, and salt, stir, and remove from the heat. Pour into six serving glasses, small bowls, or cups. Allow to cool, and then refrigerate for 6 hours.

4. To make the kumquats, combine ½ cup water and the honey in a small nonreactive saucepan. Stir well and bring the mixture to a boil, stirring regularly. Add the kumquats and return to a boil, then reduce the heat to a low simmer and cook until the kumquats are tender, about 15 to 20 minutes. Continue to cook until the liquid cooks down to a thick syrup, about 5 minutes longer. Remove from the heat and let cool to room temperature.

5. To serve, spoon the kumquats over the panna cotta.

Serves 6

bold = *foods pollinated or produced by bees*

spring

Spring is the busiest time in the hive. The warmer weather energizes the bees. By this point in the year, they have probably consumed much of the honey they have stored for the winter and need to replenish those depleted stores.

Once the temperature is above 54°F (12°C) on a consistent basis, the bees leave their hive. The queen returns to full egg production to build up her winter-depleted colony, laying up to 1,500 eggs each day.

This is a crucial time for the beekeeper. The colony might start foraging before much is in bloom. In the mountains of North Carolina, the bees could actually starve in early spring. This is when the beekeepers who took too much honey at the end of the previous season regret it. (A good rule in my area is to leave at least 50 pounds of honey in each hive for the colony.) If needed, the beekeeper can get frames of honey from another beekeeper or feed the bees sugar water, though natural beekeepers like myself try to avoid the latter as much as possible.

At this time, the beekeeper also needs to keep a close eye on his or her bees to decide whether they are outgrowing the available space in the hive. If the colony runs out of room to expand, the workers will take steps to raise a new queen. Once a new queen egg has been laid, the old queen will fly away in a swarm, taking half the colony with her and significantly slowing down the progress of the existing hive. A beekeeper must add extra hive components for the colony's expansion before the workers decide they don't have enough room.

№ 4 April

FEATURED HONEY VARIETAL:

avocado

The bee is more honored than other animals, not because she labors,
but because she labors for others. — SAINT JOHN CHRYSOSTOM

Avocado honey is made from the

nectar of avocado flowers. Avocados are grown
in the greatest numbers in California, though
they are also found in warm places like Florida
and the southwestern United States. The taste
of avocado honey, however, bears no resem-
blance to the avocado itself. The nectar of the
avocados turns into a rich, dark honey that is
reminiscent of molasses, sorghum, or Louisiana-
style cane syrup. Recipes that have been created
using any of these sweeteners, or dark brown
sugar, will work beautifully with avocado honey.

Avocado honey is made by bees at the same time
as they are pollinating citrus trees, mustard, and
other flowering trees in the growing area. The
citrus blossoms are more attractive to honeybees, having a stronger aroma. As a result, pure
avocado honey is an unusual honey, available from
hives that are located near only avocado trees.

COLOR	SMELL	TASTE	AFTERTASTE
Dark brown	Warm vegetal; hayfield	Burned sugar, molasses	Lingering cane syrup

my favorite guacamole

I *love* guacamole: My home is rarely without a couple of ripe or ripening Hass avocados, the ones with the dark rippled skin. Once the avocado skin gives to gentle pressure, the inside is ripe and ready. And if the avocado is ripe, you don't need to do much to make great guacamole. There is no honey in this recipe, but I'm including it because without bees there would be no avocados.

The ingredients:

- 2 **avocados**
- 1 small ripe tomato, diced
- 1 scallion, cut into thin rings, using white and green parts
- 2 tablespoons coarsely chopped fresh cilantro
- Zest of 1 **lime**
- Juice of 1 **lime**
- Coarse sea salt
- 1 tablespoon minced jalapeño (optional)
- 1 large bag (about 16 ounces) salted tortilla chips

Serves 4

Here's what you do:

1. Cut the avocados in half and remove the pits. (I like to nick the pit with the blade of my knife, give it a quick turn to loosen the pit from the flesh, and lift my knife blade with the pit stuck to it. A quick whack on the edge of a plastic compost pail will remove the pit.)

2. Scrape the flesh with a spoon into a medium bowl and mash with a fork, stopping while the avocado is still chunky.

3. Add the tomato, scallion, cilantro, lime zest, and lime juice. Stir gently, making sure to keep the mixture chunky. Add salt to taste. If you like things spicy, add the jalapeño.

4. Serve with the tortilla chips.

How to Prevent Browning

If you are not going to be serving the guacamole immediately, leave the avocado pit in the serving bowl to prevent the guacamole from turning brown. Just before serving, remove the pit.

avocado and mango salad

Over the years, my café has been blessed with cooks from all over the United States and even the world. This quick salad comes from El Salvador and is *so* good!

The ingredients:

- 1 **avocado**
- 1 **mango**, cut into ½-inch cubes
- Zest of 1 small **lime**
- Juice of 1 small **lime**
- 1 teaspoon finely chopped fresh **dill**
- 1 teaspoon whole fresh cilantro leaves (compost the stems)
- Sea salt
- Bibb lettuce leaves

Here's what you do:

1. Cut the avocado in half and remove the pit. Cut the flesh into ½-inch cubes while it is still in its skin. Gently remove the cubes from the skin using a small rubber spatula and place in a medium bowl. **NOTE**: If you are careful, the avocado pieces will maintain their cube shape, but if a little gets mashed, don't worry. It will still taste wonderful.

2. Add the mango cubes and season with the lime zest, lime juice, dill, and cilantro. Add salt to taste.

3. Arrange lettuce leaves on a platter, top with the avocado-mango mixture, and serve.

*Serves 4
as a side dish*

borscht with crème fraîche

Pick up a bunch of baby beets when you're at the market. Here's a nice, simple soup to make with them — a perfect spring beginning that will please even those who aren't beet lovers.

NOTE: You'll need to make the crème fraîche 24 hours in advance.

The ingredients:

FOR THE CRÈME FRAÎCHE

- 2 cups heavy cream
- 2 cups sour cream

FOR THE BORSCHT

- 1 bunch (1 pound) baby **beets**, with tops
- 1 pound new potatoes
- ½ cup **honey**, preferably avocado honey
- Sea salt
- Fresh **dill** for garnish (optional)

Here's what you do:

1. To make the crème fraîche, combine the heavy cream and sour cream in a jar. Shake thoroughly to mix and place in a warm spot overnight. (I use the top of my refrigerator.) Once the cream has thickened, refrigerate until ready to use. Crème fraîche will keep in the refrigerator for 1 week.

2. Cut off the beet tops and chop them. Combine the beets, the chopped tops, and the potatoes in a 6-quart soup pot and cover with about 2 quarts of cold water. (Don't completely fill the pot; put in just enough water to cover the vegetables.) Bring to a boil, then reduce the heat to a simmer. Cook until the beets can be pierced with a fork, about 15 to 20 minutes, depending on the size of the vegetables.

3. Prepare an ice-water bath by filling a large bowl with ice water. Remove the beets and the potatoes from the cooking liquid using a slotted spoon. Plunge the cooked beets into the ice-water bath and slip the skins off. Strain the cooking liquid through a sieve or colander into a bowl. Compost the beet skins and the cooked beet greens.

4. Grate or chop the beets and the potatoes and return to the broth. Add the honey and season with salt to taste.

5. Serve hot with 1 tablespoon crème fraîche per serving. Add a sprinkle of fresh dill on top, if you like.

Serves 4

bold = *foods pollinated or produced by bees*

the Worker Bees

To me, the workers are the most interesting bees in the colony. In many ways, they run the show. And in contrast to the queen and the drones, who each have just one main job, the workers' jobs change throughout their lives.

Worker bees, all female, start working as soon as they are born as housekeeping bees. As they emerge from their birth cell, they clean out that cell and ready it for another egg from the queen. Soon they become undertakers, helping to keep the hive clear of dead bees. From the age of 3 to 11 days, they are nurse bees, taking care of the newly laid eggs and the larvae.

From days 12 to 17, the workers first become queen attendants, tending to all the feeding and grooming needs of Her Majesty. The workers keep the queen at a constant temperature of 95°F (35°C) year-round, fanning her in the hotter months and surrounding her in a cluster in the cooler months. As they take care of her, the queen attendants take on her scent (her pheromone) and spread it to the other bees as they move about the hive.

Next, the workers transition into being house bees who greet the foragers (I'll get to them in a second), take their collected pollen and nectar, and store it in the hive. The house bees fan the watery nectar to achieve the proper consistency of honey, make and maintain honeycomb, and take on temperature-control responsibilities in the hive.

About 18 days after they are born, workers start to adopt jobs outside the hives, the first being guard bee. These bees remain by the entrance to the hive, making sure that no one who does not belong, as identified by their foreign scent, gets in.

Finally, the worker bee becomes a forager, gathering pollen and nectar and bringing it back to the hive. Scout bees make the initial forays, locating food sources and then coming back to the hive. They tell all their sisters where to go by doing a waggle dance, wiggling their bodies in a particular direction and for a very specific length of time. The direction they face indicates the direction in which their discovery is located. The length of the dance indicates the distance to the target.

glazed baby carrots

When you can pull carrots right out of the ground, put this dish on your dinner menu. If you're not the grower, try to find fresh baby carrots at your local farmers' market. The next option is the produce section of your grocery store. Use small regular carrots rather than the bagged "baby" carrots. The fresher they are, the better.

The ingredients:

- 2 bunches (about 1 pound) whole baby **carrots**, tops removed, cut into ½-inch slices
- 1 tablespoon unsalted butter
- 1 tablespoon **honey**, preferably avocado honey
- 1 tablespoon freshly squeezed **lemon juice**
- Sea salt

Here's what you do:

1. Place the carrots in a large non-reactive saucepan. Add enough water to cover the bottom of the saucepan and bring to a boil. Reduce the heat to a simmer, cover, and cook for 5 minutes or less — just long enough to cook the carrots to an al dente degree of doneness. (I like to cook them just long enough to warm them up. Baby carrots are so tender that they don't need much cooking. No one I know, except for babies, likes mushy carrots!)

2. Drain the carrots, return them to the saucepan, and add the butter. Return to the heat to melt the butter. Add the honey and lemon juice and toss to coat the carrots.

3. Add salt to taste, and serve.

Serves 4

The Importance of Nonreactive Saucepans

A nonreactive saucepan is one made of stainless steel, glass, or ceramic, or one that is coated with Teflon. The important part to note is that it will not react with acids (like lemon or vinegar). Cast iron and aluminum are examples of reactive pans.

bold = *foods pollinated or produced by bees*

rack of lamb with a coffee and avocado honey crust

This is a version of one of the staples at Blueberry Hill. My mother made a roast leg of lamb for special occasions, basting a brown sugar and mustard crust with coffee. Here's the way I like to do it, using dark avocado honey.

The ingredients:

- ¼ cup toasted fresh bread crumbs
- ¼ cup **honey**, preferably avocado honey
- ¼ cup coarse-ground **mustard**
- 1 small rack of lamb (about 2 pounds)
  Sea salt
  Coarsely ground black pepper
  Granulated **garlic**
- 1 cup strong brewed **coffee** or espresso
- ¼ cup **white wine**

Here's what you do:

1. Preheat the oven to 375°F.

2. Make a paste of the bread crumbs, honey, and mustard in a small bowl, mixing until just combined.

3. Season the lamb rack with salt, pepper, and granulated garlic. Spread the honey-mustard paste on the meat side of the rack. Place the lamb rack in a baking dish just large enough to hold it, preferably one that can also be placed directly on the stove top.

Serves 4 (2 chops per person)

Pour the coffee into the bottom of the baking dish.

4. Bake for 10 minutes, or until the crust is set, then baste with the pan drippings, trying not to dissolve the honey-mustard mixture. Continue to bake until a meat thermometer reads 130°F, 20 minutes or longer. Keep a careful watch. If your racks are small, this may take less time. Similarly, larger racks will take longer. When the proper temperature is reached, remove from the oven. Transfer the lamb rack to a warmed plate to rest while you prepare the sauce.

5. Place the baking dish on the stove. Simmer the drippings over medium-high heat and add the wine to deglaze the pan, stirring with a wooden spoon. Add any additional juices that might have collected on the plate from the resting lamb. Stir until the drippings and wine are well mixed and slightly thick, about 5 minutes.

6. Cut the rack into rib portions and serve with the warmed sauce.

Fresh Bread Crumbs

To make fresh bread crumbs, simply cut a few slices of bread into tiny pieces, or pulse a few slices in a food processor for a few seconds. You are looking for coarse crumbs, not grains of sand. To toast, toss in a dry saucepan over medium heat for a few minutes, until the bread dries and turns golden brown.

bold = *foods pollinated or produced by bees*

southern-style iced tea

I migrated to the South after a childhood in the North, where I had known iced tea as only a summertime drink. In North Carolina, tea (the "iced" is assumed) is a year-round beverage. I've gotten used to it. Here's my honeyed version. By the way, if you have enough sun in your area to make "sun tea" (by putting tea bags in a glass jar filled with water and letting it sit in the sun for a few hours), simply add honey while the water is still warm and you'll have nice sweet honey tea — just like that.

The ingredients:

6 tea bags

¼–½ cup **honey**, preferably avocado honey (depending on how sweet you like your tea)

Fresh **mint** sprigs for garnish

Here's what you do:

1. Bring 4 cups water to a boil. Add the tea bags to a heatproof pitcher, and pour the boiling water over the tea bags. While the tea is still hot, add the honey.

2. Allow to steep for 20 minutes to make a strong tea. Remove the tea bags.

3. Cool to room temperature or, better yet, make ahead of time and store in the fridge until ready to serve.

4. To serve, pour the rich sweet tea into ice-filled glasses. Garnish with mint sprigs.

Serves 4

strawberry-rhubarb cream

My mother made rhubarb cream in the spring, and I still do, pulling fresh spears from a plant I inherited when I moved into my log cabin. Rhubarb does not need bees for pollination, so I added strawberries, which do need bees, to this recipe. Strawberries and rhubarb are a wonderful combination. This recipe is like a fruit pie, but without the crust.

The ingredients:

- 1 pound rhubarb stalks, cut into ½-inch pieces
- 1 pound **strawberries**, sliced
- ¼ cup **honey**, preferably avocado honey (more or less to taste)
- 1 cup chilled whipping cream

Here's what you do:

1. Combine the rhubarb and 1 cup water in a large nonreactive saucepan. Bring to a boil, and then reduce the heat to a simmer. Cook until the rhubarb is fork-tender, stirring frequently, about 10 minutes.

2. Add ¾ pound of the strawberries and the honey, and cook for another 5 minutes. Remove from the heat.

3. Coarsely mash the fruit mixture. Add more honey to taste if the rhubarb is especially tart. Chill.

4. Whip the chilled cream until soft peaks form. Fold the whipped cream into the chilled fruit mixture. Chill for 1 hour or longer. (You could make this late in the afternoon for dinner that night.)

5. Serve in individual bowls garnished with the remaining ¼ pound strawberry slices.

Serves 6–8

---

bold = *foods pollinated or produced by bees*

raspberry

His labor is a chant,
His idleness a tune;
Oh, for a bee's experiences
Of clovers and of noon!

— EMILY DICKINSON

Raspberry honey comes from

the nectar of raspberry blossoms grown
predominantly across the northern tier of
the United States, with the exception of
the northern Plains States. The scent of the
white flowers fills the spring air with its
delicate sweetness.

Raspberry honey ranges in color from light
amber to light brown and has a mild to
medium aroma that is woodsy and caramelized. It has a lightly sweet flavor that can
be described as toffee or brown sugar. Try it
with vanilla, Champagne, chocolate, and
fresh pears or peaches.

COLOR
Light brown

SMELL
Woodsy

TASTE
Slightly sweet; brown
sugar and toffee; secondary
hint of bitterness

AFTERTASTE
Mild raspberry finish

fresh pea soup with minted cream

Mint and peas are a very nice combination, and this soup is a pretty spring starter. Instead of making a minty pea soup base, I like to put the mint in a lightly sweet, lightly lemony cream that is drizzled on top. This soup is made with fresh spring peas, though you could make it with frozen peas if you can't get the fresh kind at the market, or better yet, from your garden. Peas cook quickly, so this soup can be a spontaneous creation if you happen across fresh peas.

The ingredients:

FOR THE SOUP

- 1 tablespoon unsalted butter
- 1 leek, well washed and cut into thin strips
- 1 large sweet onion, chopped
- 2½ quarts unsalted chicken stock
- 3 cups shelled sweet peas
- 1 tablespoon salt

FOR THE MINTED CREAM

- ½ cup plain yogurt
- ½ cup sour cream
- 1 tablespoon honey, preferably raspberry honey
- Zest of 1 lemon
- Juice of 1 lemon
- 4 fresh mint leaves, minced
- 12 fresh mint leaves, for garnish

Serves 12

Here's what you do:

1. To make the soup, melt the butter in a 6-quart stockpot. Add the leek and onion and cook over low heat until transparent, about 15 minutes.

2. Add the stock and bring to a boil. As soon as the stock boils, turn the heat to low and add the peas. Cook for 4 minutes (2 minutes if you use frozen peas). Remove from the heat and allow to sit until cool. Add the salt.

3. Purée the soup in a blender in batches until very smooth. You can also use a handheld immersion blender or a food processor, but the best consistency will come from a standard blender.

4. To make the minted cream, combine the yogurt, sour cream, honey, lemon zest, and lemon juice in a small bowl. Stir well. Add the minced mint. Pulse with an immersion blender or in a blender or food processor until very smooth.

5. Ladle the soup into bowls, add 1 tablespoon of the minted cream to each bowl, and garnish with a fresh mint leaf. I like to serve it warm but not hot.

fresh baby vegetables with honeyed curry dressing

We've made this dressing for years using maple syrup, which is a fine thing, especially if you are, like me, from Vermont. But this versatile and easy dipping sauce or salad dressing also works well with honey. Give it a try. And if you have leftover sauce, slather it on a ham sandwich.

The ingredients:

FOR THE DRESSING

½ cup mayonnaise

½ cup sour cream

2 tablespoons curry powder

2 tablespoons **honey**, preferably raspberry honey

FOR THE VEGETABLES

1–2 bunches baby **carrots**, cut in diagonal slices if needed

1 bunch spring **radishes**, cut in diagonal slices if needed

1 bunch baby **turnips**, cut in diagonal slices if needed

Here's what you do:

1. To make the dressing, combine the mayonnaise, sour cream, curry powder, and honey in a small bowl and whisk. It will be easier to mix if you warm the honey (see below). Set aside in the refrigerator for an hour or so, giving the ingredients time to meld.

2. If the vegetables are not fresh from the garden and are not tiny, you may need to lightly steam them. If so, place the vegetables in a steamer basket over boiling water for 3 to 5 minutes, just until tender. Remove from heat and drain liquid.

3. Combine the carrots, radishes, and turnips in a large bowl. Add the dressing and toss to combine. Serve immediately.

Serves 6

Warming Honey

To warm honey for ease of mixing, place the honey jar in a bath of warm water for a few minutes, or run it under hot water.

---

bold = *foods pollinated or produced by bees*

fresh-from-the-garden beets with oranges and blue cheese

Pasture-raised cows eat assorted grasses that depend on bees, so I've boldfaced blue cheese in the ingredients list. Not all dairy cows are raised on pasture, though, so not all milk needs honeybees; nevertheless, there are a lot of grasses that are pollinated by bees.

The ingredients:

- 1 bunch (about 1 pound) baby **beets** (the smaller, the better), with tops
- 1 **navel orange**, peels and white membranes removed, cut into sections
- Extra-virgin olive oil
- **Red wine vinegar**
- Sea salt
- Freshly ground black pepper
- 2 tablespoons crumbled **blue cheese** (I like Maytag)
- Fresh **dill** for garnish (optional)

Here's what you do:

1. Remove the greens from the beets, cutting close to the beets. Save the greens for a garnish if you wish.

2. Set up a bamboo or stainless steel steamer, and steam the beets until a fork easily pierces the beets, about 10 minutes. (Tiny beets will cook quickly; larger ones will take longer.) Prepare an ice bath by filling a large bowl with ice water.

3. Remove the beets from the steamer and plunge into the ice bath. The skins should slip off easily (this is so much easier than peeling them).

4. Slice the beets. Arrange the beets and orange wedges on a large plate, drizzle with olive oil and vinegar, and sprinkle with salt and pepper to taste.

5. Just before serving, add the cheese. Use the reserved beet greens or dill as a garnish.

Serves 4 as an appetizer

bold = *foods pollinated or produced by bees*

seared duck breasts with raspberry honey glaze

If you don't feel like roasting a whole duck, look for individual duck breasts in the frozen-food section of your grocery store. They cook quickly and are a great, fancy-looking but simple-to-make main course. I've also had great success slicing and serving them with a little raspberry jam as a passed hors d'oeuvre.

The ingredients:

- 2 boneless, skin-on duck breasts

  Coarse salt

  Freshly ground black pepper

- 4 tablespoons **honey**, preferably raspberry honey

- ½ cup fresh **raspberries**

Here's what you do:

1. Score the duck fat using a very sharp knife, being careful not to cut all the way through the fat. Season the duck breasts with salt and pepper.

2. Heat a cast-iron skillet over high heat, and sear the duck breasts, skin side down, for 5 minutes. Turn the duck over and cook for 5 to 7 minutes longer, or until done to desired finish. (You can use an instant-read thermometer — 125°F is rare — or make a small slit on the underside of the duck to determine doneness. My preference is for medium rare, so I cook until it is light pink, or 130°F. It will continue to cook a bit after you remove it from the heat.) Transfer the duck to a warm platter.

3. Pour out the fat from the skillet, reserving 2 tablespoons. Return the skillet to the stove over medium heat and add the reserved duck fat. Add the honey and ¼ cup of the raspberries. Bring to a simmer, and allow to thicken. Season with salt and pepper to taste if needed.

4. Just before serving, slice the duck breasts and present on a platter or on individual plates and top with the raspberry sauce. Garnish with the remaining ¼ cup raspberries.

Serves 2

sweetly braised baby turnips

Spring is such a lovely time of year, especially for baby vegetables. In our local farmers' markets, it is easy to find bunches of brand-new, bite-size root vegetables: carrots, parsnips, turnips, beets. While I was never much of a turnip fan as a kid and hated them as an adolescent, I have a whole new opinion as an adult now that I have access to these sweet babies. We taught a class of first graders how to make these, along with applesauce and apple turnovers. Hands down, this was the group favorite. In this recipe, the honey glaze enhances the turnips' natural sweetness, making them a treat and a fine accompaniment to a light spring dinner.

The ingredients:

- 1 pound baby **turnips**
- 1–2 tablespoons unsalted butter
- ¼ cup **honey**, preferably raspberry honey
- ¼ teaspoon sea salt

Here's what you do:

1. If you are lucky enough to have truly fresh baby turnips, you can simply cut off the tops and the root and leave the turnip unpeeled. After a quick rinse, they'll be ready to cook. If the turnips are not right out of the ground, you may need to peel them and cut them into wedges. Keep the wedges large, about one full bite per chunk.

2. Melt the butter in a cast-iron skillet over medium heat. Add the turnips, toss, and cook, stirring often, until they can be pierced with a fork, 15 to 20 minutes. The turnips will have started to brown, which is a good thing.

3. Reduce the heat to low. Drizzle the honey over the turnips and continue to stir. Cook until the turnips caramelize and turn a chestnut color, about 10 minutes longer.

4. Right before serving, sprinkle with the salt, which will offset and enhance the sweetness of the turnips. Enjoy!

Serves 4

raspberry granita with fresh mint

Granita is a *very* simple dessert. Do not be intimidated. You don't need special machinery or fancy gadgets — only some fresh fruit, a little sweetener, a freezer, and a few hours of time. Trust me on this one. For this version, I used fresh raspberries and raspberry honey as my sweeteners.

The ingredients:

- 4 cups fresh **raspberries**
- ⅓ cup **honey**, preferably raspberry honey, plus more if needed
- Zest of 1 **lemon**
- 1 tablespoon **lemon juice**
- 6 fresh **mint** sprigs

Here's what you do:

1. Place the raspberries in a medium bowl. Pour the honey, lemon zest, and lemon juice over the berries. Allow to sit for 15 minutes.

2. Mash the berry mixture with a fork until well blended, though not puréed. Taste and add more honey if you wish.

3. Pour into a 9-inch square metal cake pan and place in the freezer, covered, for 1 hour. Remove from the freezer and stir with a fork, breaking up any frozen clumps. Put back in the freezer until firm, about 2 more hours. Remove from the freezer again and, using a chopping motion with a fork, break up the granita into flakes. Cover and return to the freezer. You can do this up to 3 days ahead of serving, if you like.

4. To serve, flake the granita once more and garnish each serving with a mint sprig.

Serves 6

bold = *foods pollinated or produced by bees*

The Drone

The drones are the only male bees in the colony. The drone has one job: to mate with a virgin queen of another colony.

Sadly, after he mates with a queen, the drone dies, leaving the reproductive portion of his body attached to his conquest. Drones do not do anything else — they do not forage, they do not collect nectar or pollen, and they do not help with housekeeping or taking care of the queen.

Once the active honey-making time is done and the temperature drops, the workers, not willing to support their non-productive brothers through the winter, kick them out of the hive, sending them to a cold death. In the spring, when the colony is once again active, the workers will allow newly hatched drones to live, as they want to make sure the genetics of their colony will be spread to other hives.

Drones are recognizable because of their size: They are smaller than the queen but bigger than the workers. Most notably, their eyes are much larger than worker bees' eyes.

If you look at a drone's eye under an electron microscope, you can see sight receptors interspersed with scent receptors that enable the drones to locate a queen when she is out on her mating flights.

strawberry fizzies

Here's a snazzy nonalcoholic cocktail for you, inspired by a gift of field-fresh berries, a table full of fresh fruit, and a friend telling me she had recently baked some strawberries in the oven. I started by cooking my fruit in a sauté pan, and it all fell into place after that.

NOTE: Blenheim ginger ale is very strong. If you are serving children or folks with less adventurous palates, you might prefer a standard ginger ale like Canada Dry or Schweppes.

The ingredients:

FOR THE STRAWBERRY PRESERVES

1 cup **strawberries**, stemmed

2 tablespoons **honey**, preferably raspberry honey

FOR EACH COCKTAIL

¼–⅓ cup **orange juice**, depending on the size of your champagne flutes

¼–⅓ cup strong ginger ale (I like Blenheim), depending on the size of your champagne flutes

1 beautiful fresh **strawberry**

1 slice **blood orange**

3 fresh **raspberries**

Here's what you do:

1. To make the preserves, combine the strawberries and honey in a small skillet. Crush the berries and stir with a wooden spoon until the mixture is very smooth. Cook over medium heat until the consistency of jam, about 5 minutes. Allow to cool.

2. To make each drink, carefully spoon 1 tablespoon of the preserves into the bottom of a champagne flute. Add the orange juice to the glass. Just before serving, fill the remainder of the glass with the ginger ale. Garnish with the strawberry, blood orange, and raspberries, and serve.

Makes enough preserves for 8 cocktails

N<u>o</u> 6 June

FEATURED HONEY VARIETAL:

tulip poplar

Seeing only what is fair,
Sipping only what is sweet,
Thou dost mock at fate and care,
Leave the chaff and take the wheat.

— RALPH WALDO EMERSON, "THE HUMBLEBEE"

Tulip poplar honey comes from

nectar collected from the flowers of tulip poplar trees, which grow in the eastern half of the United States. The fragrant flowers bloom in early spring, drawing in the bees and resulting in the first honey flows in many of these areas.

I have a huge tulip poplar tree in my backyard that fills with flowers and bees in late April and early May. The honey is a surprisingly dark color, considering how light its flavor is.

COLOR	SMELL	TASTE	AFTERTASTE
Slightly dark orange/ amber	Weak; cooked fruits	Lightly sweet; buttery finish with caramel notes	Fades away with a lingering hint of toffee

brie in puff pastry with quick strawberry preserves

Brie is a lovely cheese all on its own, but it is also great when served with a little bit of something sweet. In my café, we sometimes serve an hors d'oeuvre of a thin slice of Brie topped with a thin slice of fresh strawberry and a dab of sour cream or crème fraîche. This recipe takes those elements and fancies them up a bit. Quick preserves can be made with any seasonal fruit. If you have fruit that is starting to get too ripe, make a quick jam.

The ingredients:

FOR THE PRESERVES

- 2 pounds fresh **strawberries**, stemmed

- 2 tablespoons **honey**, preferably tulip poplar honey

- 1 tablespoon vanilla extract

FOR THE WRAPPED CHEESE

- 1 sheet frozen puff pastry

    Unbleached all-purpose flour, for the pastry

- 1 whole wheel Brie cheese (about 2 pounds)

- 1 egg

    Crackers or baguette slices, for serving

Here's what you do:

1. To make the preserves, place the strawberries in a medium, nonreactive saucepan over low heat. Stir gently if you like chunky preserves, or more emphatically if you like a smoother texture. Cook until the berries are juicy and have reached the texture you prefer.

2. Add the honey and stir until combined. Increase the heat to medium and cook until the mixture thickens slightly, about 20 minutes. Stir in the vanilla. **NOTE:** You are looking for a spreadable mixture. Stop cooking before you have fruit paste! Set the preserves aside.

3. To make the wrapped cheese, thaw the puff pastry following the instructions on the package. This will take 30 to 45 minutes, depending on the type of pastry. You should be able to unfold it without breaking it.

4. Preheat the oven to 475°F.

5. Roll out the pastry on a lightly floured surface, stretching the square until it is about 14 inches long on each side. Line a nonstick baking sheet with parchment paper and lay the pastry dough on top.

6. Place the cheese in the center of the pastry square. Spoon ½ cup of the strawberry preserves on top of the cheese. Grasp two opposite corners of the pastry square, stretch, and pull over the edge of the cheese until the corners are about 4 inches directly above the center of the cheese. Hold these corners with one hand.

7. Bring the other two corners, one at a time, up to meet the first two corners. Twist all of the corners together and arrange on the top of the cheese. You can pull the ends into fun shapes, like ribbons or leaves. Be creative!

8. Combine the egg with 1 tablespoon water. Mix well with a fork. Brush this egg wash over the entire surface of the puff pastry using a pastry brush. This will make the baked pastry shiny and much more attractive.

9. Bake for 30 minutes, or until the pastry puffs up and is a toasty golden color. If the cheese is especially cold when you start, the baking could take longer. Just look for that golden color.

10. Remove the pastry from the oven and allow to cool for at least 30 to 45 minutes before serving (unless you want a big pool of molten Brie running all over your serving table).

11. Use the parchment paper (or one or two wide, flat spatulas) to help you transfer the pastry to a serving platter. Discard the parchment.

12. Serve with crackers and get ready to accept a lot of praise from your guests.

Serves 20 as an appetizer

How Do Bees Make Honey?

The process of making honey is enough to make you shake your head in amazement. How did they ever figure all this out?

Simply put, honey is made from nectar that worker bees collect on their foraging trips. Each trip includes visits to anywhere from 150 to 1,500 flowers, all of the same kind. This is called "flower fidelity." At each flower visit, a bee sucks up a tiny amount of nectar and adds it to her "honey stomach," which is different from her "digestive stomach." As she continues to forage, a special enzyme is added to the honey stomach's contents, changing the complex sugar in the nectar into the simple sugar found in honey. By the time she is ready to return to the hive, the nectar in her honey stomach can weigh almost as much as she does, making for a low-altitude flight home.

Once back at the hive, the bee regurgitates the enzyme-enhanced nectar into a waiting honeycomb cell, one tiny drop at a time. During the busiest times of the year the nectar is transferred, mouth to mouth, from a field bee to a house bee, who will carry it to the cell and allow the field bee to go on another foraging mission.

The nectar that the field bees bring to the hive is about 80 percent water, which is way too watery. The warmth of the hive, which is always 95°F (35°C), starts to evaporate this water. The worker bees accelerate the evaporation process by fanning the liquid with their wings until the nectar is about 17 percent water — the consistency we know as honey. At that point, the workers cap the filled cell with lovely, fresh white beeswax, which keeps it safe from any spoilage. In this form, it will keep forever. Indeed, honey discovered in ancient Egyptian pyramids in modern times was still edible thousands of years later.

Bees make honey for their own nutritional needs and store it for the winter. Guidelines vary by climate, but in my area, the rule of thumb is to make sure the bees have one full medium super of honey per hive, at least 50 pounds to make it through the winter. Any additional honey can be harvested by the beekeeper.

new potato vichyssoise

If you are in a latitude similar to mine (the middle band of the country), you can find local new potatoes in the beginning of the summer. If you live elsewhere or can't locate them, look for the best, smallest potatoes you can find and you'll be fine. I like Yukon Golds, as they will lend a nice color to this soup, which is a delightful way to start an early summer dinner on the porch. Are the ladies coming for lunch? Serve them this!

The ingredients:

- ½ cup (1 stick) unsalted butter
- 6 **leeks**, well washed and cut into ½-inch slices
- 1 small sweet **onion**, cut into medium cubes (about ½ cup)
- 6 cups unsalted chicken or vegetable stock
- 3½ cups cubed new potatoes, skins left on

  Salt

  Freshly ground black pepper
- 1 cup whole milk
- 2 cups half-and-half
- ¾ cup heavy cream (optional)

  Watercress sprigs for garnish

Here's what you do:

1. Melt the butter in a large soup pot over medium heat. Add the leeks and onion and sauté until translucent, 5 to 7 minutes. Do not allow them to brown. Add the stock and the potatoes. Turn the heat to low and simmer until the potatoes are cooked, about 25 minutes.

2. Pulse the mixture using an immersion blender until it is as smooth as you wish. I recommend stopping before it is completely puréed, as some texture is a good thing.

3. Taste and add salt and pepper (the amounts will vary depending on the saltiness of your stock). Add the milk and the half-and-half and stir until well mixed. Cook over low heat for 15 minutes, making sure not to let the soup boil. Remove from the heat and allow to cool. Chill thoroughly in the refrigerator.

4. Ladle into individual bowls or cups. Drizzle with 1 tablespoon heavy cream, if you like, and add a sprig or two of fresh watercress to each bowl.

Serves 6–8

---

bold = *foods pollinated or produced by bees*

strawberry salad with honeyed balsamic vinegar reduction

This balsamic vinegar reduction keeps for a week or so, so if you have leftovers, feel free to use them to create a refreshing and sophisticated dessert: a bowl of fresh strawberries accompanied by the reduction.

The ingredients:

FOR THE BALSAMIC REDUCTION

- 1 cup balsamic vinegar
- ¼ cup **honey**, preferably tulip poplar honey

FOR THE SALAD

- 1 pound mixed spring greens
- 2 cups fresh **strawberries**, sliced in half
- ⅓ cup extra-virgin olive oil
- ½ cup crumbled feta cheese
- ½ cup slivered **almonds**, toasted (optional)

Here's what you do:

1. To make the balsamic reduction, place the balsamic vinegar in a small nonreactive saucepan. Bring to a boil, and then reduce the heat to a simmer. Cook until reduced in volume by half, about 15 minutes.

2. Add the honey and continue to cook until the mixture is a very thick syrup that easily coats the back of a spoon. (You will need to pay attention at the later stages of the cooking, as it can pretty easily get too thick, in which case you might need to thin it by adding a bit of water.) You are looking for a consistency similar to honey. Remove from the heat.

3. To make the salad, arrange the mixed greens on a serving platter. Distribute the strawberries on top of the greens.

4. Just before serving, drizzle the salad with the olive oil, then drizzle a thin stream of the balsamic reduction over the salad, making sure to get it on the berries. Scatter the cheese and the sliced almonds, if using, on top.

Serves 4

bold = *foods pollinated or produced by bees*

wild salmon with a smoky onion crust

My friend Sally comes from an island in the Pacific Northwest, where she grew up having salmon roasts on the beach, cooking the fish on cedar planks. She described her father's recipe to me, which included his secret herb concoction. Though I never tasted his version, I created my own. No beach? No worries. My version is cooked over a smoky charcoal fire in your yard.

While the salmon is delicious hot, the flavors continue to improve over time, making this an excellent next-day item or the basis for a wonderful chunky smoked-salmon salad.

The ingredients:

- 1 cup Hardwood grilling chips (Hickory, Maple, and Alderwood are good options)
- ½ cup (1 stick) unsalted butter
- 1 tablespoon Johnny's Seasoning Salt (see sidebar on following page)
- 2 tablespoons **honey**, preferably tulip poplar honey
- 4 medium **sweet onions** (for example, Vidalia or Walla Walla), sliced
- 1 side wild Alaskan salmon (about 3 pounds), skin on

Here's what you do:

1. About 1½ hours before eating, light a charcoal fire in your grill. Allow the coals to burn down to a medium heat before beginning to cook. Or, if using a gas grill, prepare a medium fire when you are ready to cook. Soak a handful of wood chips in water while you cook the onions.

2. Melt the butter in a large skillet over low heat. Add the seasoned salt, honey, and onions (you'll have a big pile of onions, but they will soften and shrink). Cook, stirring regularly, until the onions are very well cooked and resemble a thick mash almost like paste, about 45 minutes.

recipe continues on following page >

3. Place the salmon, skin side down, on the grill. You may wish to put it on a piece of aluminum foil (being careful not to cover the grill surface completely with the foil) to keep it from sticking to the grill, but if your grill is well seasoned, you can skip this step. Smear the onion mixture on top of the salmon. Close the grill cover and cook over low heat for 15 minutes.

4. Open the lid and add the soaked wood chips to the coals. This will make a smoky fire. Close the cover of the grill and allow the fish to smoke for another 15 minutes or more, depending on the thickness of the fish. To test, gently poke the fish with your finger. I like my fish on the underdone side, so I take it off the grill while it still feels soft. If you like your fish more well done, continue to cook until the meat is firmer.

5. Transfer the fish to a serving platter and loosely cover with foil. Let stand for 5 to 10 minutes, and then serve.

Serves 5–6

Seasoning Salt

If you cannot find Johnny's Seasoning Salt, you can make a reasonable facsimile by combining the following: 1 teaspoon salt, 1 teaspoon sugar, ¼ teaspoon paprika, 1 teaspoon freshly ground black pepper, and 1 teaspoon granulated garlic.

haricots verts with tulip poplar honey–mustard vinaigrette

Here's a way to cook vegetables that keeps them looking bright and fresh. You can prepare them hours before serving if you like — a convenient feature on a warm night in June.

The vinaigrette can be prepared ahead of time too. This recipe makes more than you need for the vegetables, but it is so good that you can stick the extra in your fridge for tomorrow. It is so much better than those bottled concoctions. If you can't find tulip poplar honey, meet a beekeeper in your area and get the harvest from his or her early honey flow — something light will work well.

The ingredients:

- 3 tablespoons **red wine vinegar**
- 1 tablespoon **honey**, preferably tulip poplar honey
- 1 tablespoon coarse-ground **Dijon mustard**
- 1 cup very good extra-virgin olive oil
- Sea salt
- Freshly ground black pepper
- 2 pounds fresh **haricots verts** (baby French green beans)

Here's what you do:

1. Combine the vinegar, honey, and mustard in a medium bowl and mix with a wire whisk. Drizzle the oil into the bowl in a thin stream, whisking constantly until well blended. This will cause the mixture to emulsify, so that it will stay mixed. Taste and add salt and pepper if necessary. Set aside until ready to use.

2. Bring a large pot of water to a boil. Fill a large bowl with ice water. Plunge the green beans into the boiling water, watching carefully. After about 1 minute (the time will vary depending on the age of your beans), the beans will turn a bright green. Immediately remove them from the pot and immerse them in the ice water, which will stop the cooking and keep them bright green. Once they are cool, remove them from the ice water and place them in a colander to drain.

3. Arrange the beans on a serving platter. Drizzle the vinaigrette on top of the beans, and serve.

Serves 8–10

---

bold = *foods pollinated or produced by bees*

fromage à la crème

Here's my version of one of my mother's favorites. This is an old-fashioned dessert and, like so many of that type, it is delicious. You will need one of those 15-ounce porcelain molds, often heart-shaped, with small holes in it, and some cheesecloth to line the mold. Other than that, this is a simple dish. Yes, it is rich, but you can portion out small servings. Moderation in all things allows you to have small indulgences every once in a while, right?

The ingredients:

- 8 ounces cream cheese
- ¾ cup whipping cream
- 2 tablespoons **Grand Marnier**
- 2 tablespoons **honey**, preferably tulip poplar honey, plus more for drizzling (optional)
- Fresh **strawberries** for garnish

Here's what you do:

1. Combine the cream cheese and half of the cream in a large bowl and whip until very, very fluffy. Blend in the Grand Marnier and 1 tablespoon of the honey. Whip for 1 minute longer.

2. Whip the remaining cream with the remaining 1 tablespoon honey in another bowl until stiff. Be sure to stop before you make butter!

3. Fold the whipped cream into the cheese mixture.

4. Line a 15-ounce mold with cheesecloth and pour the blended cheese and whipped cream mixture into the mold. Chill for at least 3 hours, or overnight.

5. When ready to serve, unmold onto a serving platter. Garnish with strawberries and, if you like, a drizzle of honey.

Serves 4

sparkling rosemary and melon spritzers

When you have the ripest, most fragrant melon, make this drink, especially if you are having a snazzy, seated brunch. Start or finish with this. Either way, it will generate lots of praise and requests for the recipe.

The ingredients:

1 **cantaloupe**, seeded, scooped into balls with a melon scoop

8 teaspoons **honey**, preferably tulip poplar honey

8 fresh rosemary sprigs

Sweet sparkling **wine** (or Champagne)

Fresh **raspberries** for garnish

Here's what you do:

1. About 30 minutes before your guests arrive, put 5 melon balls in each of 8 martini glasses. Drizzle about 1 teaspoon of honey into each glass and add a sprig of rosemary, piercing a melon ball with the rosemary. Allow to sit.

2. When ready to serve, fill each glass with the sparkling wine. Garnish with the raspberries. As the guests drink, the aroma of the rosemary will mix with the melon and the honey. Provide a fancy little fork or small spoon for the melon balls, if you like.

Serves 8

summer

WHAT'S HAPPENING IN THE HIVE?

Summer is a busy time for the bees. Hopefully all the preparations and attention of the spring have paid off and the bees are busily gathering pollen and nectar, raising babies, going through their worker bee cycles, and building up their honey stores for the winter.

In the area where I live, we have two major honey flows (the time when a large number of one flower or tree blooms, producing substantial amounts of one nectar, like orange blossom or clover). On each foraging flight, the bees collect nectar from just one kind of flower as long as that plant is in bloom. The beekeeper knows when these flows happen. If the desire is to collect a varietal honey, the beekeeper places the beehives near that varietal's flowers and removes the hives and extracts the honey before the bees can collect nectar from a different flower.

In my area, we have an early tulip poplar honey flow and then, after a dearth in late June, the sourwood flow happens. In warmer climates, there is a continuous series of honey flows, sometimes for the entire year.

For the beekeeper, summer is less busy than spring, though there is still important work to do. The beekeeper needs to see if the queen is doing a good job, and, for that matter, if there even *is* a queen. My mentor suggests doing a "State of the Union" examination on a nice warm day in the summer. During the midday hours, most of the bees are out of the hive on foraging flights and the bees inside the hive are busy dealing with the incoming deliveries of nectar and pollen. With many of the bees either out or busy, it's possible to take a look without creating too much disturbance.

FEATURED HONEY VARIETAL:

sourwood

Pleasant words are as an honeycomb, sweet to the soul, and health to the bones. — PROVERBS 16:24

Sourwood honey is a very special

honey from my region; it is with pride that I write about it. In fact, the sourwood honey gathered by a local beekeeper, Virginia Webb, has been awarded the designation of "Best Honey in the World" at competitions in France and Ireland.

This unusual and rare honey is made from the nectar of the sourwood tree, which has a short bloom time. In June and July, the 40- to 50-foot-high trees fill with flowers that, from a distance, look like an elegant, lacy lady's glove. Tiny white flowers resembling lily of the valley decorate the roadsides from southern Pennsylvania to northern Georgia.

Sourwood honey's distinctive characteristics are its spicy aroma and flavor and tangy aftertaste. The color can vary and darkens over time, but the honey is absolutely worthy of its reputation as the best in the world.

COLOR
Varies from extra light to light amber

SMELL
Extremely aromatic, with hints of cinnamon and cloves

TASTE
Pleasantly and not overwhelmingly sweet; spicy, sometimes described as resembling baking spices with hints of clove and nutmeg

AFTERTASTE
The flavor lingers on the sides of the tongue

dolce e forte (sweet and strong)

In my travels to Tuscany, I have had the good fortune to taste some very old Tuscan dishes based on recipes collected by my friend there, a scholar of Tuscan foods. This one has stayed with me since I first tasted it, and I give it a hearty two thumbs up.

I love the sweet and salty combination here. Make the mixture in advance and stick it in the fridge. An hour or so before serving, spread the mixture on baguette slices. You may serve as is, or arrange on a baking sheet for toasting. Then, as your guests arrive, place the baking sheet under the broiler and toast lightly.

The ingredients:

- 1 cup pancetta, roughly chopped
- 1 cup pine nuts
- 1 cup golden **raisins**
- ½ cup brined capers, drained
- ½ cup candied **orange peel** (optional)
- 4 tablespoons butter
- 1 teaspoon unbleached all-purpose flour
- 2 tablespoons **honey**, preferably sourwood honey
- 6 tablespoons distilled **white vinegar**
- Baguette, sliced diagonally ½ inch thick

Serves 8 as an appetizer

Here's what you do:

1. Combine the pancetta, pine nuts, raisins, capers, and candied peel, if using, in a food grinder or food processor. Pulse a few times until the mixture is the consistency of ham salad. Be careful not to make it too fine.

2. Combine the butter, flour, and honey in a medium skillet. Cook over medium heat until it starts to foam.

3. Add 3 tablespoons water and the vinegar to the skillet. Cook for 2 minutes to unite the ingredients. Stir in the pancetta mixture and cook for just 1 minute longer. It will be a thick, spreadable mixture.

4. Spread on the sliced baguette. Serve as is. Or, if you like, preheat the broiler. Place the slices on a baking sheet and broil for 1 to 2 minutes, just until it toasts.

bold = *foods pollinated or produced by bees*

How is Honey Harvested?

First, the beekeeper needs to remove the honey-filled super from the hive. This can be a tricky process because the bees, understandably, become agitated when part of their home is dismantled. Smoking them gently while maintaining a calm demeanor can be helpful.

Once the super has been removed, it is taken to the "honey house," where it will be harvested. The honey house can be a garage, a kitchen, or even a dedicated shed — anywhere that can be closed off to the bees, who, upon smelling the honey, will be eager to take it back to their hives.

To harvest the honey, the beekeeper must remove the wax cappings on the cells, either with a heated knife or with a capping tool that simply scratches the surface of the capped honey. A plain table fork works just fine too, but a heated knife is easier to use when you're doing a number of frames, as it cleanly removes just the cappings. The cappings are set aside for another use, like candle-making, at a later date.

In the simplest, lowest-tech version, the uncapped frames are left to drip into a waiting container, with gravity doing the work. This is a very slow process, however, so most beekeepers prefer to use a mechanical extractor. The most basic extractor is just a barrel that holds the uncapped frames. It can be turned quickly by hand or electronically, allowing centrifugal force to spin the honey out of the frames. Smaller home-use extractors hold four to six frames, larger ones hold nine, and commercial ones hold many more. When the honey from the frames has been spun out, the frames are removed and filled frames are put in their place. This is repeated until all the frames are emptied.

The extracted honey collects in the bottom of the extractor, pours out from a spigot, and is filtered to make sure that little bee parts and beeswax shreds don't end up in the honey. The honey is then put in jars. Ta-da!

chilled cucumber soup

This odd recipe is one of my mother's, transformed by honey. A perfect soup for a hot summer night, it is delicious, cooling, and quite easy. Do not try this soup unless you have really fresh cucumbers.

The ingredients:

- 1 (12-ounce) bottle light-colored beer (not a "lite" beer but something like a lager)
- ½ cup sour cream
- 2 medium **cucumbers**
- 1 teaspoon **honey**, preferably sourwood honey
- 1 teaspoon sea salt
- ½ teaspoon granulated **garlic**

Here's what you do:

1. Slowly pour the beer over the sour cream in a large bowl.

2. If the cucumbers are very, very fresh, you can leave them unpeeled, but if they are not right from your garden or local farmers' market, peel them. Cut the cucumbers into a very small dice (reserve 4 thin slices for garnishing the soup) and add them to the sour cream mixture.

3. Slightly warm the honey by running the jar under warm water for about 1 minute. Add the salt, garlic, and warmed honey to the bowl. Stir well to combine. If you prefer a smoother texture, pulse in a blender or, easier still, use an immersion blender. Chill in the refrigerator.

4. Ladle into chilled soup bowls or cups. Garnish with a thin slice of cucumber and be prepared to hear some "oohs" and "aahs."

Serves 4

bold = *foods pollinated or produced by bees*

laurey's sweet potato salad with sourwood honey

This salad came about when one of our cooks accidentally overcooked some potatoes as he was trying to make sweet potato chips for me. On his way to the trash he absentmindedly ate one, found its taste intriguing, and decided to play with it, turning it into one of the most popular items we serve.

For the most reliable result, use a convection oven, as you want the sugars to almost burn, and that doesn't happen nearly as well with a conventional oven.

The ingredients:

- 2 pounds sweet potatoes
- 1 medium **red onion**
- ½ cup extra-virgin olive oil
- 1 teaspoon granulated **garlic**
- 1 teaspoon kosher salt
- 1 teaspoon freshly ground black pepper
- 1 tablespoon **honey**, preferably sourwood honey
- Juice from ½ **lemon**
- ⅓ cup chopped fresh **parsley**

Here's what you do:

1. Preheat a convection oven to 350°F or a conventional oven to 375°F.

2. Peel and quarter the sweet potatoes and red onion. Slice approximately ⅛ inch thick in a food processor or on a mandoline.

3. Mix the potatoes and onion in a large bowl. Toss with about ¼ cup of the olive oil (enough to coat the potatoes, not more), and the garlic, salt, and pepper.

4. Line a baking sheet with parchment paper. Spread the mixture on the baking sheet. Bake, turning occasionally, until the edges of the potatoes are browned, but be careful not to burn them. This will take about 35 minutes in a convection oven and a little longer in a conventional oven.

5. Remove from the oven and allow to cool for 10 minutes.

6. Whisk the honey, lemon juice, and the remaining ¼ cup olive oil in a small bowl. Add to the potato mixture and toss to combine.

7. Add the parsley and serve. We usually serve this at room temperature. I've heard of people having it hot, but I can't say I've ever tried it.

Serves 6

Move Over, Idaho

Guess what? We grow more sweet potatoes in North Carolina than anywhere else in the entire world!

vermont-style summer squash casserole

This is a perfect side dish in midsummer, when squash, tomatoes, and sweet onions are abundant. Be sure to use really sharp or extra-sharp Vermont cheddar! (Okay, if you have a good sharp local cheese, I'll let this slide, but make sure it's sharp, and then rename the dish using that state's name, just to keep it all on the up and up.) You may bake this casserole ahead of time and then warm it up before you are going to serve it. The flavors will have melded by then, and you will be quite pleased with the results.

The ingredients:

- 2 **yellow squash**, sliced into thin rounds (about 2 cups)
- 2 **zucchini**, sliced into thin rounds (about 2 cups)
- 1 medium **sweet onion**, sliced into thin rounds
- 2–3 medium tomatoes, sliced into thin rounds
- ½ teaspoon coarse salt
- ¼ teaspoon freshly ground black pepper
- ¼ teaspoon granulated **garlic**
- 2 cups shredded sharp Vermont cheddar cheese

Here's what you do:

1. Preheat the oven to 350°F.

2. Butter the bottom and sides of a broiler-proof casserole dish. Arrange a layer of half of the yellow squash on the bottom, followed by a layer of half the zucchini, onion, and tomatoes. Season with half of the salt, pepper, and garlic. Cover with a layer of half of the cheese. Follow with a second layer of the remaining yellow squash, zucchini, onions, tomatoes, and seasoning before topping with a second layer of the remaining cheese.

3. Bake for 30 to 35 minutes, until the edges of the dish are bubbling. Increase the heat to broil and, watching carefully, broil until the cheese is browned (not burned!) and bubbly. Serve warm.

Serves 6–8 as a side dish

bold = *foods pollinated or produced by bees*

grilled garlic shrimp with a fresh heirloom tomato sauce

When you find fresh shrimp, try this preparation and gather the neighbors for a light summer meal. The sauce is also a fine accompaniment to grilled chicken, scallops, or a fresh fish fillet. Best of all, you can prepare the sauce ahead of time if you like.

The ingredients:

FOR THE MARINATED SHRIMP

- ½ cup extra-virgin olive oil
- ¼ cup **red wine vinegar**
- 2 **garlic** cloves, minced
- 36 large shrimp, peeled and deveined
- 12 (6-inch) wooden skewers

FOR THE TOMATO SAUCE

- 3 pounds assorted large heirloom tomatoes
- 1 small sweet **onion**, minced
- 1 **garlic** clove, minced
- ½ teaspoon sea salt
- ¼ teaspoon freshly ground black pepper
- 1 tablespoon **honey**, preferably sourwood honey

  Sea salt

  Freshly ground black pepper

- ¼ cup fresh basil leaves, cut into thin strips (*chiffonade* is the formal name for this cut), plus more for garnish

Here's what you do:

1. To marinate the shrimp, combine the olive oil, red wine vinegar, and garlic in a large bowl. Stir to combine. Add the shrimp and allow to sit, covered, for 1 hour. Stir occasionally.

2. Prepare a medium fire in a charcoal or gas grill. Soak the skewers in water for at least 30 minutes to prevent them from burning.

3. To make the tomato sauce, bring a medium pot of water to a boil. Place one or two tomatoes at a time into the boiling water. Watch them and, as you see the skin split, remove with a slotted spoon and place in a bowl of cool water. At this point, it will be very easy to slip off the skins.

recipe continues on following page >

4. Cut the peeled tomatoes into a small dice. Put the cut tomatoes into a large bowl. Add the onion, garlic, salt, pepper, and honey. Stir gently to combine.

5. Skewer the shrimp, 3 per skewer. Grill the skewered shrimp for 1 to 2 minutes on each side, until they are pink. Sprinkle lightly with sea salt and a couple of grinds of fresh pepper.

6. Just before serving, add the basil leaves to the tomato sauce. Taste and add more salt if necessary.

7. Ladle the tomato sauce onto a serving platter and arrange the skewers on top of the sauce. Garnish with more basil leaves and enjoy!

Serves 6

capturing a swarm

A beekeeper can capture a resting swarm relatively easily with a bit of planning, as long as the swarm is not perched at the top of a very tall tree. On the ground underneath the swarm, the beekeeper places an empty "hive body" (the bottom box of the hive) on top of a sheet. Into the hive body goes a frame of brood (eggs and larvae) from another hive, along with nine other empty frames.

The beekeeper shakes the branch or object that is harboring the swarm. With luck, the bees and their queen will fall directly into the waiting box. Bees who miss the box will fall onto the sheet and walk right into their new home, guided by the pheromones of the queen and the bees on the brood frame.

After dark, once all the bees are tucked into their new home, the beekeeper covers the entrance of the hive, picks up the captured swarm in its new hive box, and takes it to the home apiary, where he has the start of a new colony.

switchel

Switchel is a curious beverage that is also known as Haymaker's Punch. Back when haying was a very energetic occupation, farm wives would bring gallons of Switchel to the men working in the fields. Popularized by Vermont doctor D. C. Jarvis in his best-selling 1958 book *Folk Medicine: A Vermont Doctor's Guide to Good Health*, this honey and apple cider vinegar drink became a must-have beverage.

This is a great thirst quencher and is very simple to make. Much healthier and more effective than sugary sports drinks and easy enough to regularly keep on the top shelf of your refrigerator, Switchel might just become your favorite go-to summer or post-exercise drink. The spiciness of the sourwood honey makes it extra special.

The ingredients:

- ½ cup **honey**, preferably sourwood honey
- ½ cup **apple cider vinegar**
- 1 tablespoon sliced fresh gingerroot (optional)

Here's what you do:

1. Pour 2 quarts of water into a large pitcher. Combine the honey and apple cider vinegar in a bowl and stir well to combine. If either ingredient is cold, you might need to warm them slightly or the honey will not mix well. Add the mixture to the water. If you like a sweeter drink, add more honey. If you prefer a less sweet version, add a bit more vinegar.

2. If you're a ginger fancier, add the gingerroot slices. Keep the pitcher in your refrigerator for a hot day.

Serves 8–10

broiled peaches

In my humble opinion, not much beats a fat, juicy Georgia peach. Once when I was in high school, my sister, brother-in-law, and I drove from Illinois to Florida. I remember the trip especially well, since I had just received my driver's license and drove for much of the trip. There was an accident on a side road in Georgia that stopped traffic, and a local peach stand owner took advantage of the opportunity, strolling up and down the road and selling the fattest peaches I had ever seen. It was absolutely impossible to eat one without ending up with a shirt full of juice. No matter — the mess was so worth it.

When you want to dress them up a bit (not that you need to with a perfect peach), try this recipe. It also works well with grapefruit or pears. The fruit needs to be fully ripe, ready to eat. Serve the broiled fruit for a different, light first course, or offer it for dessert with a scoop of vanilla ice cream or a dollop of vanilla yogurt.

The ingredients:

- 5 large ripe freestone **peaches**
- 2 tablespoons butter
- 2 tablespoons **honey**, preferably sourwood honey
- ¼ cup plain yogurt

Here's what you do:

1. Preheat the broiler to high.

2. Cut the peaches in half lengthwise and remove the pits. Place cut side up in a shallow broiler-proof baking dish that is the right size to hold the fruit halves without having them rolling all over the place. If necessary, shave a tiny bit off the bottom to help them stand up straight.

3. Combine the butter and honey in a small bowl and microwave on high for about 25 seconds, or until the butter is melted. Pour the honey butter over the peach halves. If not baking right away, cover and refrigerate.

4. Place the peaches under the broiler and watch carefully. Broil until the butter melts and bubbles, 3 to 5 minutes or longer, depending on the temperature of the peaches when you start to broil. You want the honey and butter to brown but not burn.

5. Spoon a small amount of yogurt on each peach half and drizzle juices from the pan on top. Serve immediately.

Serves 10

bold *= foods pollinated or produced by bees*

blueberry

The men of experiment are like the ant, they only collect and use; the reasoners resemble spiders, who make cobwebs out of their own substance. But the bee takes the middle course: it gathers its material from the flowers of the garden and of the field, but transforms and digests it by a power of its own. — FRANCIS BACON, *FIRST BOOK OF APHORISMS*

Blueberry crops are grown

commercially across the middle and north-eastern sections of the United States, most notably in Maine and New Jersey. It is estimated that the blueberry crop increases by 1,000 pounds of fruit *per acre* when the bushes are pollinated by honeybees.

A logical match for recipes with summer fruits, blueberry honey is a light, smooth, easy-tasting option when you want an unassertive sweetener.

COLOR
Light to medium amber

SMELL
Medium-strength aroma

TASTE
Buttery, with a hint of toasted almond

AFTERTASTE
Lingering; sweet butter

bbt bites

I love a good BLT sandwich. August is the perfect time to serve these little BBT (Bacon, Basil, and Tomato) Bites: The tomatoes are ripe, the basil is plentiful, and it's easy and quick to prepare. I've made and served this appetizer in non-tomato times too, but as you'd imagine, it is much better with sun-warmed beauties, especially if they come from your garden.

This recipe will make more basil cream than you need. It keeps nicely and is delicious with grilled fish, turkey sandwiches, and omelets.

The ingredients:

FOR THE BBT BITES

- 2 slices bacon
- 1 large tomato (Cherokee Purples are particularly delicious)
  Coarse sea salt
  Freshly ground black pepper
- 4–5 ½-inch-thick slices whole-grain bread

FOR THE BASIL CREAM

- ½ cup fresh basil leaves, plus more for garnish
- ¼ cup mayonnaise
- ¼ cup sour cream
- ¼ teaspoon honey, preferably blueberry honey
  Sea salt (optional)

Here's what you do:

1. Preheat the oven to 375°F.

2. To make the BBT bites, place a wire rack on top of a baking sheet and arrange the bacon on the rack. Bake for 10 to 15 minutes, until it is nicely crisp. (The bacon fat will drip onto the baking sheet and the bacon will remain crisp.) Remove from the oven and allow to cool. Cut the bacon into ½-inch pieces.

3. Slice the tomato into ⅛-inch-thick slices and then cut the slices into 1-inch squares. And yes, it is okay if you don't make exact squares. Sprinkle lightly with salt and pepper.

4. Cut the bread into 2-inch squares.

5. To make the basil cream, combine the basil, mayonnaise, sour cream, and honey in a food processor. Pulse until well blended. Taste and add salt if needed.

6. To assemble, spread the basil cream onto half of the bread squares. Add the tomato slices. Add a dot more of basil cream and top with the bacon squares. Garnish with fresh basil leaves, place the tops on the sandwiches, and pass to your guests!

Serves 8
as an appetizer

broiled tomatoes with seven seasonings

August is certainly tomato time and, if you get tired of tomato sandwiches (though that would never happen to me), let me suggest this quick vegetable accompaniment. If you don't have great tomatoes, you can try this with the hydroponic ones, but the fresher the tomato, the better the finished product. This is one of my mother's recipes. I *love* it and trust you will too.

NOTE: If the tomatoes are less than garden-ripe, bake them for 10 minutes at 325°F before broiling, which will help soften and cook the tomatoes and allow the seasonings to meld.

The ingredients:

1 large tomato

Kosher salt

Freshly ground black pepper

Dried basil

Granulated **garlic**

⅛ teaspoon **honey**, preferably blueberry honey

2 thin rounds Bermuda **onion** (about the same diameter as the tomato)

1 tablespoon grated Parmesan cheese

Here's what you do:

1. Cut the tomato in half through the "equator" and place, cut side up, on a baking sheet.

2. Sprinkle a generous pinch each of salt, pepper, basil, and garlic on top of each tomato half. Drizzle the honey on top.

3. Place 1 onion slice on top of the seasonings on each tomato half. Add the Parmesan.

4. Broil, watching closely, for 5 minutes, or until the Parmesan is toasted. Serve immediately.

Serves 2

chunky heirloom tomato gazpacho with lemon crème fraîche

I love tomatoes served in almost any fashion. I am one of those people who plants extra tomatoes just so I can pick them as an after-work snack, hot off the vine. For me, there can never be too many tomatoes. If you're looking for a bit of a twist on a standard gazpacho, give this a look. I'm enchanted by all the colors in heirloom tomatoes, and this recipe shows them off. Topped with a chilled crème fraîche, which is also very easy to make, you'll make your friends very happy. Trust me on this one.

NOTE: You'll need to make the crème fraîche a day in advance. The gazpacho may also be made a day in advance, if you wish.

The ingredients:

FOR THE CRÈME FRAÎCHE

- 2 cups sour cream
- 2 cups heavy cream

FOR THE GAZPACHO

- 3 pounds varied heirloom tomatoes, diced
- 2 Kirby **cucumbers**, diced
- 1 **garlic** clove, minced
- ½ cup extra-virgin olive oil
- ¼ cup **red wine vinegar**
- 2 tablespoons **Worcestershire sauce***
- Sea salt (optional)
- Freshly ground black pepper (optional)
- 1 **lemon**
- Fresh cilantro leaves, for garnish

Here's what you do:

1. To make the crème fraîche, combine the sour cream and the heavy cream in a small container with a lid and stir to mix. Let sit in a warm place, like on top of your stove, overnight. You'll have some leftover crème fraîche; it will keep in the refrigerator for a week or so.

2. To make the gazpacho, combine the tomatoes, cucumbers, and garlic in a large bowl. Drizzle with the olive oil, red wine vinegar, and Worcestershire. If you like a chunky style, stir gently to combine. If you prefer a smoother version, pulse the mixture in a food processor or blender, being careful not to blend too much, so that you preserve the individual colors of the tomatoes. Taste and add salt and pepper if needed.

3. Allow to sit for at least 1 hour before serving. It will keep very well in your refrigerator and gets more delicious as it sits.

4. To serve, spoon the gazpacho into individual bowls. Drizzle some crème fraîche on top, then add a squeeze of fresh lemon juice and a couple of leaves of cilantro.

Serves 6

* honeybees pollinate tamarind

bold = *foods pollinated or produced by bees*

barbecued triggerfish with laurey's grilling sauce

You may never have had triggerfish. Until my local fish shop got some, I confess I hadn't either. If you can find it, it's worth trying. The fish comes from the North Carolina coast, has a buttery flavor, and holds up under a broiler or, with gentle handling, on the grill. If you can't find it, substitute a thick white fillet, like line-caught Atlantic cod or Pacific halibut. Since you are grilling, a thin fillet will not hold up well.

The sauce is my amalgamation of a few tomato-based barbecue sauces I've had and liked. It's also great on grilled chicken, with the charred flavor of the grill mingling well with the sweetness and saltiness of the sauce. It will keep for a week in the fridge or a month or so in the freezer.

The ingredients:

- 4 tablespoons extra-virgin olive oil
- 1 garlic clove, thinly sliced
- 2 small sweet onions, thinly sliced
- ½ teaspoon sea salt
- ¼ teaspoon freshly ground black pepper
- ½ teaspoon Colman's dry mustard
- 1 cup ketchup
- ¼ cup apple cider vinegar
- 2 tablespoons honey, preferably blueberry honey
- 2 tablespoons lemon juice
- 2 tablespoons soy sauce
- 1 tablespoon Worcestershire sauce*
- 2 pounds North Carolina triggerfish or another thick fillet, like cod, skin on

* honeybees pollinate tamarind

Here's what you do:

1. Heat the olive oil in a medium saucepan over medium-low heat. Add the garlic and onions and sauté until the slices brown slightly, about 5 minutes.

2. Reduce the heat to low. Add the salt, pepper, dry mustard, ketchup, vinegar, honey, lemon juice, soy sauce, and Worcestershire. Cook for about 20 minutes. You are looking for a thick, brushable consistency, so keep cooking if your sauce is too thin. Conversely, if it is too thick, add a bit of water. Keep the sauce warm over very low heat until ready to serve.

3. Place the fish in a large baking dish. Brush each side with a generous amount of the barbecue sauce. Allow the sauce to soak in while you are heating the grill.

4. Prepare a medium fire in a charcoal or gas grill.

5. Place the fish skin side down directly on the grill. Cook for 3 to 5 minutes, or until the skin doesn't stick to the grill, then gently turn the fish over and grill the other side for 2 to 3 minutes. Your actual cooking time will depend on the thickness of the fillet and the heat of your coals. Turn over once more and brush the top of the grilled fish with more barbecue sauce.

6. Transfer from the grill to a serving platter. Serve with additional warm sauce.

Serves 4

sweet fact

A honeybee flies 15 miles per hour and gathers pollen and nectar in a 2- to 5-mile radius from the hive.

watermelon salad

This is a very pretty salad, especially when arranged on a platter for a buffet. I've also served it individually plated as a great first course. Either way, it is light, fresh, and delicious. The contrast of the sweet melon and the salty olives and cheese is surprising and oddly perfect.

The ingredients:

- 4 cups **watermelon** cut into large chunks
- ½ cup crumbled feta cheese
- ½ cup pitted Kalamata olives
- 2 tablespoons extra-virgin olive oil
- Fresh **mint** leaves

Here's what you do:

For a buffet presentation, just before serving, arrange the watermelon chunks on a large, flat platter. Top with the cheese and olives. Drizzle with the olive oil and garnish with the mint. For individual plates, just make smaller versions the same way.

Serves 4

bold = *foods pollinated or produced by bees*

fresh fruit shrub

Here's another old-fashioned fruit drink along the lines of Switchel (page 117). This one is a tiny bit more complicated, but once you make the basic concentrate, you can add sparkling water anytime you like for a refreshing drink.

Use whatever berries are fresh at the market. Blueberries, raspberries, and blackberries work well.

The ingredients:

- 4 cups fresh assorted **berries**
- 2 cups **apple cider vinegar**
- 1½–2 cups **honey**, preferably blueberry honey
- Sparkling water

Here's what you do:

1. In a large glass jar, combine the berries and vinegar. Cover and let sit in your refrigerator for 3 days.

2. If you want a smooth drink, strain the liquid through a colander into a bowl to remove the solid parts of the fruits. If you prefer a chunkier drink, don't strain; just mash the fruits together with the liquid, leaving the fruit solids in the mix.

3. Pour the fruity vinegar into a medium, nonreactive saucepan. Add the honey and stir well to combine. Bring to a boil, and cook for 2 minutes. Remove from the heat and allow to cool. Pour the sweetened mixture back into the glass jar and return to the fridge until ready to use.

4. When you'd like to have a glass of the drink, fill a glass with ice, add ¼ cup of the fruit mixture, and fill the remainder of the glass with sparkling water. Taste and adjust with more concentrate if needed. You can also make this by the pitcher. Just figure out how many glasses your pitcher holds and portion accordingly.

Makes enough concentrate for 10–15 servings

sautéed fruit compote with brandy

My mother served a fruit compote in wine every night during the summer at Blueberry Hill. As I was looking at her recipe, I was surprised to see that it included only canned fruit. She wrote her books in the early 1960s. My, how things change! If she were still around, she would be searching out the freshest local fruits.

My recipe was inspired by the memory of those compotes and also influenced by apothecary jars filled with brandied fruits I have seen and tasted. When the fruit is fresh, buy it and follow this simple method for a spectacularly simple dessert in the months to follow. Of course, I recommend using fruits that need honeybees.

The ingredients:

- ½ pound **blackberries**
- ½ pound **blueberries**
- ½ pound **cherries**, stemmed and pitted
- ½ pound fresh **figs**, cut in half or, if small, left whole
- ½ pound **peaches**, pitted and sliced
- ½ pound **plums**, pitted and sliced
- 1–2 cups **honey**, preferably blueberry honey (amount will vary depending on tartness of the fruit)
- 3–4 cups good-quality brandy (or more, depending on the size and shape of your jar)
- Vanilla ice cream for serving

Here's what you do:

1. Combine the blackberries, blueberries, cherries, figs, peaches, plums, and honey in a large bowl. Stir gently to combine without damaging the fruit.

2. Gently spoon the fruit into a 1-gallon jar with a lid or an apothecary jar with a lid. Add enough of the brandy to cover the fruit.

3. Allow to sit in the back of your refrigerator or in a cool place for at least 1 month, though no one is going to fault you for dipping in earlier than that. You can also add more fruit over time; just make sure it is covered with more brandy.

4. If you like, quickly flambé before serving for a showy finish. To flambé, warm the fruit and its liquid in a saucepan over medium-high heat. If you're not serving it soon after making it, you may need to add some fresh brandy to get a good flame. Bring to a simmer. Lift out a spoonful of some of the liquid and, working over the saucepan, carefully light it with a match. Put the flaming liquid back into the saucepan. Along with adding a dramatic flourish to your meal, this will burn off some of the alcohol in the liquid. Make sure to gather your guests around before the show!

5. To serve, spoon over vanilla ice cream.

Serves 20 or more

bold = *foods pollinated or produced by bees*

sage

For so work the honey-bees, creatures that by a rule in nature teach,
The act of order to a peopled kingdom. — WILLIAM SHAKESPEARE, HENRY V

Sage honey comes in three

common types: purple sage, black button sage, and white sage. Not surprisingly, sage honey comes predominantly from California, where the dry climate provides the perfect growing requirements.

If you have the chance to visit a farmers' market and can meet a sage honey producer, fill your basket with it. The light color suggests a modest flavor, but in reality, the subtle floral beginning opens up to reveal a sweet clover taste with a lingering rose and light violet aftertaste.

On one of our first honey tastings, I was delighted to find how popular sage honey was. Phineas, a young fellow who helped test some of our recipes, chose this as his favorite from all the 12 honeys he tried. How about that?

COLOR	SMELL	TASTE	AFTERTASTE
Very light wheat	Mildly floral	Sweet; rose petals and delicate flowers	More very light flowers

Healthy Honey and Other Products

Written accounts of the medicinal benefits of honey go back to the time of Hippocrates and before. Whole books have been written on the subject of apitherapy, but let me give you an overview.

Honey contains protein, vitamins, minerals, and enzymes, and so is an excellent addition to a healthy diet. It has been found to help combat sleep problems, digestive issues, sore throats, colds, and much more. When applied externally, honey has been shown to be one of the best burn therapies there is. It is antimicrobial and so helps with scar reduction and a long list of other problems.

Propolis, the resinous "glue" that keeps the inside of the hive safe from disease, is a highly effective antiseptic. It contains antioxidants and has anti-inflammatory properties. Some people find that mixing propolis with hot water makes a powerfully effective sore throat remedy.

Bee pollen, the high-protein granules that the bees store and use for their own nutrition, contains all of the essential amino acids, vitamins, and minerals that we need. It helps with allergies and is an effective addition to cancer treatments, among other uses.

Bee venom, the material that is injected when a bee stings, is anecdotally reported to be helpful in treating arthritis. In fact, many beekeepers say they have been cured of arthritic hands and joints after being stung in the course of working with their bees, since the bees often sting right on an arthritic joint.

Royal jelly, the hugely potent food that is fed to all bees for the first three days and then to the queen eggs for their entire gestation, has been used by practitioners of Chinese medicine for centuries. The list of its benefits is long and enormously inclusive.

Beeswax, the material that bees use to build honeycomb, is used in many body treatments.

There is much more to say about the important subject of apitherapy than I have the room to do here. If you want to delve further, do visit the website of the American Apitherapy Society.

butternut squash soup with toasted almond brittle

Here's a quick soup that is very satisfying. You'd think that a soup like this would have to cook for hours, but the entire recipe, from the time you think of it until the time you serve it, can be done in about 45 minutes.

I found a recipe for a microwaveable brittle, which makes this a very reasonable thing to tackle. I must say that as we were testing the recipe, we created a near inferno in the microwave. But rest assured that we've tweaked it to perfection so that you can proceed with confidence.

The ingredients:

FOR THE SOUP

- 2 **butternut squash**
- 1 tablespoon **sunflower oil**
- 3 cups unsalted chicken (or vegetable) stock
- ⅓ cup **honey**, preferably sage honey
- ⅓ cup half-and-half
- 1 teaspoon salt

FOR THE BRITTLE

- 1 cup **slivered almonds**
- 1 cup sugar
- ¼ cup **honey**, preferably sage honey
- ¼ cup light corn syrup
- ⅛ teaspoon salt
- 1 tablespoon butter
- 1 teaspoon vanilla extract
- 1 teaspoon baking soda

Here's what you do:

1. Preheat the oven to 350°F.

2. To make the soup, cut the squash in half and remove the seeds. Place, cut side down, on a baking sheet. Spray or brush with the oil.

3. Bake for 30 minutes, or until very tender. Remove from the oven and allow to cool.

4. Scrape the pulp into a medium soup pot. Add the chicken stock and honey. Using a handheld blender, purée until very smooth. Add the half-and-half and blend. Add salt to taste. Warm over low heat, but do not boil.

5. To make the brittle, combine the almonds, sugar, honey, corn syrup, and salt in a microwaveable bowl. Microwave on high for 2 minutes. Stir the mixture. Microwave for 2 minutes longer. Add the butter and microwave for 1 minute longer. Stir in the vanilla and baking soda.

6. Line a baking sheet with parchment paper. Pour brittle onto baking sheet and allow to cool, about 30 minutes.

7. Break the brittle into small pieces and sprinkle over the warm soup. You will have extra brittle for snacking.

Serves 8–10

apple and celery salad with sage honey vinaigrette

This is a deliciously crunchy salad that has sweet, salty, and sour tastes — three of the "big four" (which means that it will make your guests' palates very happy). It has both soft and crisp textures, too, so you get both a flavor and a texture blast. The idea for this salad was given to me by a snazzy bride whose wedding we catered a few years ago. It is now part of our permanent repertoire.

The ingredients:

FOR THE SALAD

2 stalks **celery**, thinly sliced on the diagonal

1 tart **apple** (Granny Smiths work well, but any local, green apple will be fine), unpeeled and thinly sliced

   Lemon juice (optional)

FOR THE DRESSING

¼ cup **apple cider vinegar**

1 tablespoon **honey**, preferably sage honey

½ cup **sunflower oil**

   Salt

4 tablespoons shaved Parmesan or Grana Padano cheese

Here's what you do:

1. Combine the celery and apple slices in a medium bowl. It is best to cut the apples right before serving, but if not serving immediately, toss the mixture with lemon juice to prevent the apples from turning brown.

2. To make the dressing, combine the vinegar and honey in a small bowl and whisk to mix. Drizzle the oil into the bowl in a thin stream, whisking constantly until well blended. Add a pinch of salt.

3. Place ½ cup of the apple-celery mixture in the center of each salad plate. Drizzle with the vinaigrette and sprinkle with 1 tablespoon of shaved Parmesan.

Serves 4

Shaved Cheese

You can use a vegetable peeler to shave a block of Parmesan or Grana Padano cheese.

bold = *foods pollinated or produced by bees*

california-style baby back ribs with sage honey

This is my version of a favorite recipe of my mother's. We had helpers from all over the world at Blueberry Hill. This recipe originally came from Susie Kitigawa, a young Japanese student who helped my mother cook and gave us all our first taste of the Far East. Susie used white sugar. I prefer this version, which features sage honey. I like to serve this dish as an appetizer.

The ingredients:

- 1 cup tamari
- ½ cup **honey**, preferably sage honey
- 4 **garlic** cloves, peeled
- 1 whole rack baby back ribs (about 2 pounds)
- 1 small **Meyer lemon**

Here's what you do:

1. Preheat the oven to 325°F.

2. Combine the tamari, honey, and garlic cloves in a small saucepan over medium-high heat. Bring to a boil, and then reduce the heat to a simmer. Cook for 5 minutes, but watch carefully so the mixture does not burn. Remove from the heat and set aside.

3. Meanwhile, place the rack of ribs in a deep baking pan. Add 1 cup water, or enough to cover the bottom of the pan. Cover with aluminum foil and place in the oven. Bake the ribs for 45 minutes.

4. Remove the water from the baking pan and brush the marinade on the ribs. Bake, uncovered, basting frequently with the marinade, for 30 minutes longer, or until thoroughly cooked.

5. To bring a crisp finish to the ribs, preheat the broiler or prepare a hot fire in a charcoal or gas grill. Broil or grill the ribs just until crispy, watching carefully to prevent burning.

6. Squeeze the Meyer lemon over the ribs just before serving. Cool, cut, enjoy!

Serves 4 (3 ribs per person) as an appetizer

bold = *foods pollinated or produced by bees*

charleston chicken country captain

Here's a recipe that is popular around the southern parts of the country, especially the closer one gets to the coast. It is a colorful dish, and it is pleasantly flavorful without being too spicy. We serve it a lot for casual dinners; it also makes a great fall lunch dish. Make some rice to serve with it, add a salad, and you'll be set.

The ingredients:

½ cup extra-virgin olive oil

1 stick butter

4 whole boneless, skinless chicken breasts (about 2 pounds total), cut into large chunks (about 6 pieces per half-breast)

2 stalks celery, sliced on the diagonal

1 small red onion, sliced into thin rounds

1 red bell pepper, seeded and cut into thin wedges

1 yellow bell pepper, seeded and cut into thin wedges

1 tablespoon curry powder

3 medium tomatoes or one 16-ounce can whole peeled tomatoes, cut into large cubes

2 cups golden raisins

4 tablespoons honey, preferably sage honey

1 (13.5-ounce) can coconut milk

2 cups sweetened flaked or shredded coconut, toasted

1 cup slivered almonds

2 tablespoons orange juice

Here's what you do:

1. Heat the oil and 2 tablespoons of the butter in a large skillet over medium heat. Sauté the chicken pieces, a few at a time, until browned on all sides. Remove seared pieces from the pan and place in a 9- by 13-inch casserole dish. Continue to cook until all of the chicken is done.

2. Melt 2 tablespoons of the butter in the same skillet over medium heat. Sauté the celery, onion, and bell peppers until the onion is transparent, about 5 minutes. Stir in the curry powder, cook until combined, about 1 minute longer, and remove from the heat.

3. Preheat the oven to 350°F.

4. Place the tomato chunks on top of the seared chicken. Sprinkle the raisins on top of the chicken and then cover with the sautéed pepper and onion mixture. Drizzle with 2 tablespoons of the honey. Carefully pour the coconut milk into the dish, trying not to pour it on top of the chicken pieces but evenly distributing the liquid around the bottom of the casserole.

5. Bake for 20 minutes, or until you can see bubbles forming around the edge of the baking dish.

6. Mix the coconut and the almonds in a separate bowl. Spread on top of the chicken and vegetable mixture.

7. Combine the remaining 4 tablespoons of butter and the honey in a microwaveable bowl and microwave on high for 25 seconds. Add the orange juice and mix. Drizzle over the coconut-almond topping.

8. Bake the chicken for another 10 minutes, or until the crust is lightly toasted. Serve warm.

Serves 12

Toasted Coconut

To toast coconut, simply toss it in a skillet over medium heat for about 5 minutes, or until lightly toasted.

elsie's stewed apples

This applesauce recipe is a tiny bit more involved than a recipe for a traditional applesauce, but the extra step of sautéing the apples is worth it. Use tart local apples; avoid apples that have a mealy texture and that are overly sweet. The best thing to do is to go to your farmers' market or local produce stand and taste as many apples as you can, choosing the one that is most appealing (no pun intended).

The ingredients:

2–3 tablespoons unsalted butter

2 pounds **apples**, peeled, cored, and cut into thick slices

½ cup **honey**, preferably sage honey

½ cup **white wine**

Zest of 1 **lemon**

Juice of 1 **lemon**

Fresh sage sprigs for garnish (optional)

Serves 6–8

Here's what you do:

1. Melt the butter in a large skillet over medium heat. Add the apples, turn the heat to high, and sauté until they begin to brown on the edges, about 5 minutes. If some are getting too well done, remove them and place on a plate while the rest continue to cook, then return them to the skillet when all are done.

2. Reduce the heat to low and add the honey, wine, lemon zest, lemon juice, and ½ cup water. Cover and allow to cook until the apples are tender but still firm; you don't want them to turn into applesauce!

3. Serve this in a bowl, tuck in a couple of sprigs of fresh sage from your garden, and tell your guests all about sage honey.

bold = *foods pollinated or produced by bees*

rosemary lemonade with sage honey

This recipe started with my friend Chris, who shared many of her favorite recipes from her travels in Mexico. This is one of my favorites, though her version was made with sugar. The honey here adds a layer of complexity that makes the flavor so much more interesting. We now serve her not-too-sweet rosemary lemonade in my café on a regular basis, varying the type of honey according to availability. (And yes, sometimes we still use sugar, but that's just when honey is scarce.)

The ingredients:

½ cup **honey**, preferably sage honey

16 fresh rosemary sprigs

12 **lemons** (to make about 2 cups juice)

Here's what you do:

1. To make a simple syrup, combine the honey and ½ cup water in a small saucepan and bring to a boil over high heat.

2. Turn off the heat. While the syrup is still hot, add 4 of the rosemary sprigs and let steep in the syrup until the syrup is at room temperature. Remove the rosemary and place in your compost pile.

3. Squeeze the lemons. Combine the lemon juice with the warm rosemary syrup. Add 8 cups water to the mixture.

4. To serve, fill each glass with ice and then pour the mixture into each glass. Add a sprig of the remaining fresh rosemary to each glass.

Serves 12

apple-honey-nut "thing"

Guests at our Inn often sent my mother their favorite recipes. I'm not sure where this one came from, but it quickly became one of our favorites. My sisters and I named this dessert, which is not really a cake and not really a pie, a "thing." It is still a favorite with all of us. This recipe makes two "things"; you may freeze the second one if you like.

The ingredients:

FOR THE "THING"

- 1 cup cake flour, sifted
- 4 teaspoons baking powder
- 1 teaspoon salt
- 4 eggs
- 1 cup **honey**, preferably sage honey, at room temperature
- 2 cups tart, local **apples**, peeled and cut into ½-inch cubes
- 2 cups chopped **pecans**
- 2 teaspoons vanilla extract

FOR THE CREAM

- 1 cup whipping cream
- 2 teaspoons **honey**, preferably sage honey
- 1 teaspoon **sherry**

Here's what you do:

1. Preheat the oven to 350°F. Grease two 10-inch pie plates with butter.

2. To make the "thing," sift the flour, baking powder, and salt together in a small bowl.

3. Cream the eggs and honey in a large bowl with a handheld mixer. Add the sifted flour mixture, apples, pecans, and vanilla. Carefully mix until well blended.

4. Pour the batter into the prepared pie plates. Bake for 30 minutes, or until the top is light brown and has a spongy look and a toothpick comes out dry when inserted in the center. Transfer to a rack and allow to cool.

5. To make the cream, whip the cream in a bowl using a handheld mixer on high. When soft peaks are just starting to form, add the honey and sherry. Continue to whip until soft peaks form.

6. To serve, cut the "thing" into wedges and top each wedge with a hearty spoonful of the honey cream.

Makes 2 pies; serves 12

fall

Fall and winter are important times for the beekeeper.
In the beginning of fall, when honey can be removed from the hives
following the final honey flow, it is crucial that the beekeeper check in on
the bees and, if necessary, help the colony. It is very important to have a
strong colony at this time, especially in places with hard winters.

The beekeeper should leave enough honey for the bees at the end of the
harvest to keep them going during the winter. Guidelines vary according
to climate, but in my area, the rule of thumb is to leave at least one medium
super full of honey, 50 to 60 pounds, in each hive. If the bees run out of
honey in the winter, a beekeeper can feed the bees with sugar or fondant
(a sugary paste), but honey is, by far, the best food.

The placement of the honey in the hive is also crucial. When the colony
shrinks and clusters around the queen as the outside temperatures drop, the
worker bees will not travel very far to find food — even two frames away
— so it important that the honey be nearby. Before the temperatures drop
for the winter, the beekeeper might need to shift things around, moving the
honey stores as close to the cluster as possible. If the beekeeper waits too
long to examine the hive, he or she will expose the bees to the cold when
lifting the hive cover.

Beekeeping is a delicate and challenging hobby, but to me there are few
things as rewarding as coming home to hives full of healthy bees who have
successfully made it through an entire year.

eucalyptus

Eucalyptus honey varies from

light amber to very dark brown, depending
on where the eucalyptus is growing. It has a
stronger taste than the lighter honeys, but
that is very pleasing to those folks who have
a more adventurous palate.

In our guided honey tastings, we set up two or
three stations, each with three or four honeys.
All of these honeys are mild- to medium-flavored,
approachable, and enjoyable to most palates. Off
to the side of the room, we have an "adventure
table," which is where the hard-core tasters
end up. At this table, we sample the dark, deeply
aromatic, assertively flavored honeys. Eucalyptus
honey is not the strongest honey around, but
we place it at this table. To me, the recipes in
this chapter, most containing this dark honey, are
perfect for the chillier days of autumn.

COLOR	SMELL	TASTE	AFTERTASTE
Very dark brown	Strong; woody; smoky; resinous	Sweet propolis; pine resin	Lingering, slightly bitter sweetness

fried green tomatoes with gremolata cream

This is a good appetizer for fall and a terrific way to use up the green tomatoes that might still be hanging around in your garden. The farmers' markets in my area often sell green tomatoes toward the end of the season, and your grocery store might very well have them too. Of course, by now you know that I believe local produce tastes better.

Gremolata is an Italian flavoring mixture that works very well in many recipes. Consisting of lemon zest, minced parsley, and grated Parmesan, it is one of those magical combinations whose sum is delightfully much more than any of its components.

The ingredients:

FOR THE GREMOLATA CREAM

Zest of 1 **lemon**

2 tablespoons minced fresh **parsley**

2 tablespoons grated Parmesan

½ cup sour cream

FOR THE TOMATOES

2 large green tomatoes

1 cup buttermilk

½ cup coarsely ground cornmeal

½ cup unbleached all-purpose flour

1 teaspoon salt

1 teaspoon freshly ground black pepper

Vegetable oil for frying (I like peanut oil, but canola or corn oil will work too)

Serves 4 as an appetizer

Here's what you do:

1. To make the gremolata cream, combine the lemon zest, parsley, and Parmesan in a small bowl. Add the sour cream, stir, and set aside.

2. To make the tomatoes, slice the tomatoes into ¼-inch-thick slices. Soak in the buttermilk for 5 to 10 minutes.

3. Mix the cornmeal, flour, salt, and pepper in a shallow dish. Dredge the buttermilk-soaked slices in the flour mixture.

4. Heat ½ inch of oil in the bottom of a large skillet over medium-high heat. Fry the tomato slices until golden brown, about 2 minutes. Turn over and fry on the other side until golden, about 2 minutes longer.

5. Drain the slices on paper towels. Allow to cool slightly, and then serve immediately, accompanied by the gremolata cream. If the slices are too large, cut them in half, thirds, or quarters.

bold = *foods pollinated or produced by bees*

Mind Your Beeswax!

After honey, beeswax is the most well-known product of honeybees. We use it for candles, lip balm, salves, and potions, but it is of critical importance to the bees.

Bees use wax to construct honeycomb, which is where the brood is raised and honey is stored. When bees establish a new colony out in the wild, the first thing they do is build their home using their own wax. When bees are given hives by a beekeeper, the beekeeper will frequently insert a thin sheet of hexagon-imprinted beeswax in each frame, giving the bees a foundation on which to build the honeycomb. (Bees eat their own honey to get the energy to make wax, consuming 6 to 8 pounds of honey to make 1 pound of wax, so giving them a head start can be a good thing for the beekeeper.)

The cells of the honeycomb are constructed from tiny flakes of wax, which come from a gland in the nurse bees' abdomens. When the wax comes out of the nurse bees' glands, it is clear. The bees transfer the flakes from their hind legs to their front legs and then into their mouths. They chew the wax flakes, which softens them and makes them malleable enough to form into honey-comb. This new beeswax is snowy white and is a beautiful sight and smell for a beekeeper.

Bees construct perfect hexagonal cells on both sides of the foundation sheet. Scientists have determined that hexagonal cells are an extremely stable shape for storing the most honey using the least amount of wax.

Interestingly, the cells are tilted slightly upward on each side of the frames so that the nectar and honey do not drip out. And, as you may recall (see page 41), the fully drawn honeycomb of one frame is ⅜ inch away from the fully drawn comb of the adjoining frame, respecting the "bee space" at which the bees are most comfortable.

It's all quite complicated and intriguing, like almost everything else about bees.

yukon gold potato chowder

As a gal who grew up in New England, I love a bowl of thick chowder, be it clam or potato. I remember once in college slipping on a rock and actually falling into the ocean in November, getting completely soaked. Luckily I was wearing a thick red wool sweater that kept me warm. On the way back to my dorm to change, we passed a chowder house that beckoned, even though I was still wet. That hot bowl of heaven kept me going until I made it back to dry clothes.

Though I am putting this recipe in the fall section, I tested it on a warm day in May and it was instantly devoured by all my friends. Don't feel that you have to wait for the cold temperatures.

The ingredients:

- 3 slices bacon, cut into 1-inch pieces
- 1 large **carrot**, peeled and cut into ½-inch cubes
- 1 sweet **onion**, cut into ½-inch cubes
- 3 **garlic** cloves, sliced
- 6 tablespoons butter
- 3 bay leaves
- 1 tablespoon **Worcestershire sauce***
- 1 teaspoon salt
- 3 pounds Yukon Gold potatoes, cut into ½-inch cubes
- 3 quarts rich chicken or vegetable stock
- ¼ cup unbleached all-purpose flour
- 1 cup heavy cream
- Fresh **dill** sprigs for garnish

Here's what you do:

1. Sauté the bacon pieces in a 6-quart stockpot over medium heat for about 5 minutes, stirring to keep them from sticking.

* honeybees pollinate tamarind

2. Add the carrot, onion, garlic, and 2 tablespoons of the butter, and cook until the onion becomes transparent, about 5 minutes longer. Add the bay leaves, Worcestershire, and salt.

3. Add the potatoes and stock. Cook over medium heat until the potatoes are tender, 30 to 45 minutes. Turn off the heat and allow to sit.

4. Melt the remaining 4 tablespoons butter in a small saucepan over low heat. Add the flour and stir to make a smooth paste.

5. Gradually add the heavy cream to the paste, stirring constantly until it is thoroughly combined and very smooth. Add 1 cup of the broth from the soup to the cream mixture, continuing to stir constantly until very smooth. Add the cream-broth mixture to the soup, stirring well.

6. Warm the soup slowly, stirring regularly, over medium heat, until it has thickened to a creamy consistency.

7. To serve, ladle soup into bowls and garnish each bowl with a sprig or two of fresh dill.

Serves 12

bold = *foods pollinated or produced by bees*

roasted delicata squash with tuscan kale

This salad was inspired by my friend Annie, a wonderful cook who introduced me to delicata squash. The squash is naturally sweet and pairs so nicely with the kale and the other tastes of Italy and the Mediterranean. This recipe calls for pine nuts, which are quite expensive these days, but the buttery texture and flavor is so delicious that I am reluctant to suggest an alternative. This dish is great as a lunch salad or as a warm side dish with the Deviled Beef Bones (page 156).

The ingredients:

- 3 **delicata squash** (about 3 pounds total)
- Extra-virgin olive oil
- ¼ teaspoon coarse salt
- ¼ teaspoon freshly ground black pepper
- 1 pound bow-tie pasta
- 2 bunches Italian (lacinato) **kale**
- ½ cup pine nuts
- 1 cup crumbled feta cheese

Here's what you do:

1. Preheat the oven to 350°F.

2. Cut the squash in half lengthwise and remove the seeds. Cut into 1-inch chunks (there's no need to remove the edible skin). Arrange on a baking sheet and spray or brush with olive oil. Season with the salt and pepper. Roast 10 to 15 minutes, until tender. Allow to cool.

3. Fill a large pot with water, add salt, and bring to a boil over high heat. Add the pasta and cook until just tender. Drain, rinse, and set aside.

4. Remove the leaves of the kale from the stems and cut into large pieces. Set up a steaming basket over boiling water, and steam the kale just until bright green, about 2 minutes. Remove and plunge into ice water to stop the cooking and keep them bright green.

5. Toast the pine nuts in a small dry skillet over medium-low heat until light brown, 3 to 5 minutes. (Stay nearby while you're toasting. Left unattended, they can easily burn.)

6. Combine the pasta, kale, squash, and pine nuts in a large bowl. Toss, and then add the cheese. Taste and adjust the seasonings. Enjoy!

Serves 8–10

bold = *foods pollinated or produced by bees*

deviled beef bones

I grew up with these wonderful beef bones, which appear in the Leftovers section of my mother's *Blueberry Hill Cookbook*. We served a lot of standing rib roast beef at Blueberry Hill. After my father carved the meat off the roast, my mother stuck the cooked rib bones in the freezer until she had enough to make a whole dinner of them. The fat rib bones still had a lot of meat, enough that one or two were sufficient for a meal. These days, if you don't have any leftovers around, know that you can buy uncooked beef bones at a good butcher shop or meat department. Ask the butcher to remove the tough membrane that holds the bones together.

The barbecue sauce is a dark, wintery mixture featuring eucalyptus honey, which is one of my favorites. It resembles molasses or Louisiana cane syrup and, combined with the mustards, is a delicious sauce for beef bones.

The ingredients:

- 1 teaspoon **dry mustard**
- ½ teaspoon salt
- ¼ cup **Dijon mustard**
- 1 tablespoon **apple cider vinegar**
- 1 tablespoon **honey**, preferably eucalyptus honey
- 1 tablespoon molasses
- 1 tablespoon **Worcestershire sauce***
- 6–8 whole beef rib bones, cooked (see below for cooking instructions if using uncooked ribs)

* honeybees pollinate tamarind

Here's what you do:

1. Preheat the oven to 425°F if using cooked ribs, or 450°F if using uncooked ribs.

2. Combine the dry mustard, salt, Dijon mustard, vinegar, honey, molasses, and Worcestershire in a small bowl. Whisk well.

3. If your ribs are already cooked, place them on a baking sheet, brush with the barbecue sauce, and cook in the oven for 15 to 20 minutes. Finish them under the broiler for 5 to 7 minutes, until crispy.

4. If your ribs are not cooked, place them in a baking pan, brush with the barbecue sauce, and bake in the oven for 15 minutes. Remove from the oven and reduce the heat to 350°F. Brush the ribs again with the sauce and return to the oven for 20 to 25 minutes longer. Remove the ribs once more and brush with more sauce. Turn the heat to broil and broil for 5 to 7 minutes, until the ribs are crispy (but not burned!). Serve warm.

Serves 3–4

braised fall greens with apples

At our farmers' markets, the farmers sell bags of mixed braising greens, like kale, collards, mustard greens, and turnip greens. This is a fun way to try different greens without having to commit to a whole mass of one kind. And since it is fall, why not add a bit of fruit?

The traditional Southern way to cook greens is to boil them with seasonings, reduce the heat, and cook for hours. A hunk of fatty bacon makes them tasty, too. But I find that they don't need to cook that long to be good. And the bacon, though flavorful, is also not needed. Try mine and see what you think. Around here, people tease me and call them "Yankee greens," but they eat them and are happy just the same.

The ingredients:

- 2 tablespoons extra-virgin olive oil
- 1 medium sweet **onion**, cut into large chunks
- 2 **garlic** cloves, sliced
- ¼ cup **apple cider vinegar**, plus more for finish
- 2 tablespoons **honey**, preferably eucalyptus honey, plus more for finish
- 2 tart **apples**, cored, unpeeled, and cut into chunks
- 3 pounds mixed **braising greens**, cut into large pieces, thick stems removed

  Sea salt

  Freshly ground black pepper

Here's what you do:

1. Heat the oil in a large skillet over medium heat. Add the onion and garlic and sauté until they start to brown, about 5 minutes.

2. Add ½ cup water, and the vinegar, honey, and apples, stirring to combine. Add the greens and stir, folding in the uncooked greens on the top as the ones in the bottom of the pan wilt, until all the greens have been incorporated.

3. Lower the heat to a simmer, cover, and cook for 5 to 30 minutes, depending on the age of the greens. If the greens are young, they will only need to cook for 5 minutes. Older greens, like a mix of mostly collards, will need to cook longer.

4. Taste and add salt and pepper to your liking. A pinch of salt and a grind or two of pepper will do the trick. Too much more will mask the sweet flavor of the greens, so do taste them first, okay?

5. Splash with a bit more vinegar and one more drizzle of honey, and serve.

Serves 6–8

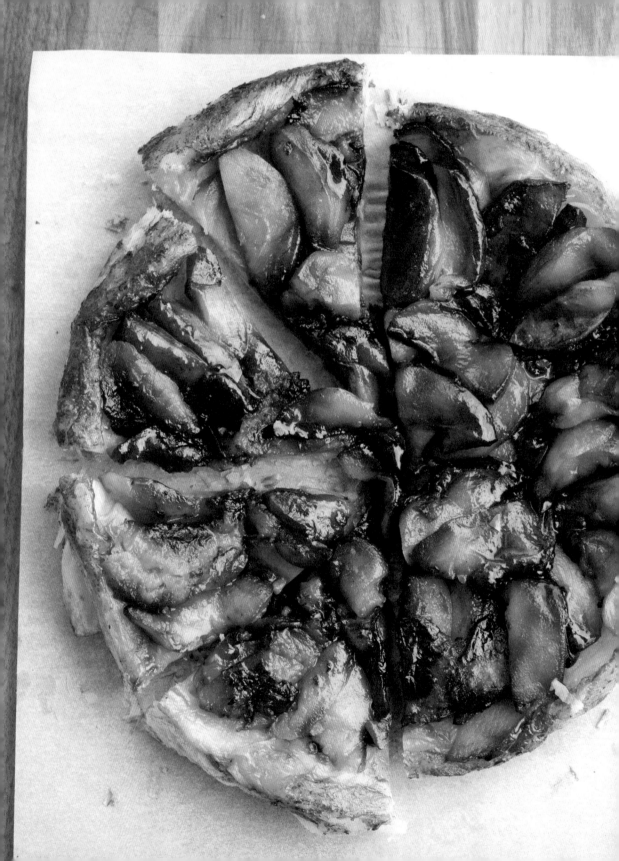

easy tarte tatin

I love tarte tatin, the inverted apple pastry, but I am not the best baker in the world, as I'm not really patient with careful measuring. Frankly, I am much more comfortable cooking than baking. But this recipe will produce a grand result even if you're not a serious baker. And if you have any leftovers, they make a great breakfast.

Get the best tart local apples you can find. With that start, you'll do very well.

The ingredients:

- 1 sheet frozen puff pastry
- ½ cup (1 stick) unsalted butter
- ½ cup **honey**, preferably eucalyptus or local honey
- 3 pounds tart **apples**, peeled, cored, and sliced into wedges

  Unbleached all-purpose flour, for the pastry

  Ice cream for serving (optional)

Here's what you do:

1. Following the instructions on the package, thaw the puff pastry. This will take 30 to 45 minutes, depending on the type of pastry. You should be able to unfold it without breaking. Set aside.

2. Melt the butter in a 10-inch cast-iron skillet over medium-low heat. Add the honey. Stir well to blend thoroughly. Carefully arrange the apple wedges in the bottom of the skillet in a decorative pattern, taking special care on the first layer, as it will end up being the top of the tart. Take care, also, to fill in any holes with other apple pieces. Continue to layer the apples until you have used all the apple slices. Since they will shrink as they cook, you want the uncooked apples to be higher than the edges of the skillet, so add more if needed.

3. Cook over medium heat on the stove until the juices bubble up and change from clear to a rich amber color, 15 to 25 minutes, depending on the heat and the consistency of your apples. As they cook, press the apples down with a rubber spatula; once the juices are visible, baste the apples with the juices. Keep an eye on them and don't allow them to burn. Remove from the heat and allow to cool slightly.

4. Preheat the oven to 475°F.

recipe continues on following page >

5. Roll out the thawed pastry on a floured surface, until it is a square that can comfortably fit over the skillet. Lay the puff pastry over the cooked fruit, making sure that the pastry completely covers the apples. Tuck the pastry into the sides of the skillet, sealing in the apples.

6. Bake the pastry-covered skillet in the oven for 20 to 30 minutes, or until the pastry puffs up and turns a golden brown. Remove from the oven and allow to cool completely.

7. Place a serving platter on top of the cooked pastry and, holding tight, flip the skillet over so that the tart comes out of the skillet and ends up on the platter, pastry side down. Remove any of the cooked apples that might have stuck to the skillet and tuck them into the tart as needed.

8. Serve with ice cream, if you like, though it is perfect just as it is.

Serves 6–8

sweet fact

Honeybees must visit 2 million flowers to collect enough nectar to make 1 pound of honey. It takes 55,000 miles of flight to reach this many flowers — a little more than twice the Earth's circumference.

russian tea

Chai first comes to mind when I think of a spicy beverage, but hot Russian tea is a fine alternative. In Russia, they brew a strong sweet spiced tea and then, separately, add more hot water. Guests serve themselves, adding water as they prefer. In this recipe, all are mixed together before serving, but you can share the story of what the Russians do when you serve it.

Russian tea is a great beverage for a chilly day. And of course, making it with honey is much healthier than using sugar or jam, as is more commonly done.

The ingredients:

- ¾ cup **honey**, preferably eucalyptus honey, plus more if needed
- 2 sticks **cinnamon**
- 1 teaspoon whole cloves
- 1 **lemon** rind, cut into quarters
- 1 **orange** rind, cut into quarters
- 6–8 Earl Grey tea bags
- Juice of 6 **lemons**
- Juice of 3 **oranges**
- Fresh **mint** leaves for garnish

Here's what you do:

1. Combine the honey, cinnamon, cloves, lemon rind, and orange rind in a large stockpot. Add 2 cups water and cook over medium heat for 10 minutes. Remove from heat and allow to sit for 1 hour.

2. Bring 1 gallon water to a boil in a separate large stockpot. Turn off the heat and add the tea bags. Allow to steep for 5 minutes. Remove the tea bags and add the spice mixture, the lemon juice, and the orange juice.

3. Taste for sweetness and add more honey if needed. Serve hot or chilled. Garnish with mint leaves.

Serves 15–20

№ 11 November

FEATURED HONEY VARIETAL:

cranberry

The only reason for making a buzzing-noise that I know of is because you're a bee [. . .] And the only reason for being a bee that I know of is making honey [. . .] And the only reason for making honey is so as I can eat it.

— WINNIE THE POOH IN A. A. MILNE'S *THE HOUSE AT POOH CORNER*

Cranberries may often be thought of as coming from bogs in Massachusetts, but the truth is that all of the Northeast, some of the Great Lakes states, and some states in the far Northwest also grow these tart berries. Cranberries grow in the wetlands. At harvest time, the bogs are filled with water. The berries are raked and float to the surface, making a beautiful bright red scene in the fall.

Cranberry honey is fruity, providing a wonderful pairing with fall fruits and anything you might like to cook with those fruits.

COLOR	SMELL	TASTE	AFTERTASTE
Medium reddish brown	Fruity	Warm with cooked fruit hints; figs, dates, and apricots	None

cranberry chutney

This is my personal favorite contribution to a Thanksgiving meal. And, in truth, to me it is not really Thanksgiving without it. I'll eat the stuff squished from a can, but this, chunky and real, wins. You can make chutney at other times of the year with fresh peaches, apples, pears, or figs — or whatever else happens to be ripe in that season. Depending on the sweetness of the fruit, you may need to adjust the amount of honey you add.

The ingredients:

1 (1-pound) bag fresh **cranberries**

2 **navel oranges**, unpeeled, cut into 6 wedges and then into thin slices

⅓ cup **golden raisins**

½ teaspoon ground **cinnamon**

½ teaspoon whole cloves

½ teaspoon ground **ginger**

½ teaspoon kosher salt

½ cup **apple cider**

½ cup **honey**, preferably cranberry honey

¼ cup **apple cider vinegar**

Makes 3 cups

Here's what you do:

1. Pour the cranberries into a 2-quart pot. Add the oranges, raisins, cinnamon, cloves, ginger, salt, apple cider, honey, and vinegar. Bring to a boil, and then reduce the heat to low and simmer until the chutney thickens, about 20 minutes.

2. Remove from the heat and serve warm or, if you prefer, chill and serve cold.

baked acorn squash with cranberry honey

Truly one of my most favorite recipes for squash, this version replaces the customary maple syrup that my mother used with cranberry honey. I'm giving you a recipe for one squash, which will serve two people. Expand as you need, depending on the number of guests you are expecting.

The ingredients:

- 1 **acorn squash**
- ½ cup half-and-half
- 1 tablespoon **honey**, preferably cranberry honey
- 1 tablespoon butter

Here's what you do:

1. Preheat the oven to 325°F.

2. Cut the squash in half through the stem end. Trim off a small piece from the bottom of each half to stabilize the squash, but be careful not to cut too deeply or all the liquid will run out. Scrape the seeds from each half.

3. Fill each half of squash three-fourths full with the half-and-half. Divide the honey and butter between each squash half. **NOTE:** Do not overfill, as the liquid will boil over and make it difficult to pick up the cooked squash.

4. Bake for 45 to 50 minutes, until the squash is browned around the edges and the liquids have cooked into a bubbly custard. Serve hot.

Serves 2

candy roaster squash soup with frizzled shallots

You're lucky you if you live in a place where you can buy Candy Roaster squash fresh from a farmer. This funny-looking squash is mildly sweet but has a rich, old-fashioned taste similar to, but better than, butternut squash. Adding a little honey brings out its sweetness in a wonderful way.

I am fortunate to have access to locally grown Candy Roasters. In fact, one of the farmers who provides produce to my restaurant was instrumental in saving the seeds of this very squash, keeping it from extinction. If your farmers don't have any, ask them to grow some for you or, better yet, grow some yourself (you can purchase the squash seeds from small seed companies like Sow True Seeds). If all else fails, use butternut squash.

The ingredients:

1 **Candy Roaster squash**

¼ cup plus 3 tablespoons extra-virgin olive oil

¼ teaspoon sea salt, plus more if needed

¼ teaspoon freshly ground black pepper

2 sweet medium **onions**, cut into chunks

4 **garlic** cloves, sliced

¼ cup **honey**, preferably cranberry honey, plus more if needed

2–3 cups half-and-half

1 cup sliced shallots

Here's what you do:

1. Preheat the oven to 350°F.

2. Cut the squash into quarters. Scrape out the seeds (save them to grow!) and brush the insides of the squash with 1 tablespoon of the olive oil. Sprinkle with the salt and pepper.

3. Place skin side up on a baking sheet and bake for 30 to 40 minutes, until soft. Remove from the oven and allow to cool enough to touch comfortably.

4. Scoop out 4 cups of squash pulp and set aside. Freeze the extra and use it the next time. Put the empty peel in the compost pile.

5. Heat ¼ cup of the olive oil in a medium soup pot over low heat. Add the onions and garlic and cook gently until transparent, about 5 minutes. Add the squash pulp and stir to combine. Cook for 5 minutes.

6. Add the honey and stir well. Slowly add the half-and-half, stirring continuously. Taste and adjust the seasonings. You may need a bit more honey or a bit more salt, depending on how sweet your squash is. If you want a smooth soup, pulse in your food processor.

7. Heat the remaining 2 tablespoons olive oil in a small skillet over medium-high heat and sauté the shallots until crisp, 5 to 7 minutes. Cover a plate with paper towels and place the frizzled shallots on top. Season lightly with salt.

8. Sprinkle the crisp shallots on top of the soup just before presenting to your guests. If you happen to have another uncooked squash, put it right on the table as a centerpiece. *That'll* certainly get the conversation going.

Serves 6–8

sweet fact

Bees make honey from the flowers and flowering plants that are within a 3 to 5 mile radius of their hive. If you suffer from spring allergies, try eating honey from your area, as this is known to alleviate allergy symptoms.

california endive with pomegranate seeds and shaved parmesan

The endive is the forced bulb of the chicory root, another vegetable that needs honeybees for pollination (check out various websites for the fascinating story of how endives are produced).

Have you seen the beautiful California endive with its luscious red-tipped leaves? You can use the Belgian kind, of course, but look and ask for the California-grown ones. This is a lovely salad for an early winter meal.

Serves 6–8

The ingredients:

- 2 tablespoons **honey**, preferably cranberry honey
- 2 tablespoons **apple cider vinegar**
- ¼ cup extra-virgin olive oil
- ½ teaspoon sea salt
- Freshly ground black pepper
- 4–5 heads **endive**
- Seeds (arils) from ½ **pomegranate**
- 1 cup shaved Parmesan (about ¼ pound)

Here's what you do:

1. Combine the honey and vinegar in a small bowl. Drizzle the oil into the bowl in a thin stream, whisking constantly until well blended. Season with the salt and a couple of grinds of pepper. Taste and adjust the seasonings.

2. Cut off the root end of the endive heads and separate into individual leaves. Arrange the leaves on individual salad plates. Sprinkle with the pomegranate seeds and drizzle with the vinaigrette. Scatter about 1 tablespoon of Parmesan shavings on each serving. Alternatively, cut the spears into small, diagonal pieces, place in a large salad bowl, and toss with the pomegranate seeds, vinaigrette, and Parmesan.

Shaved Parmesan

It is easy to shave your own Parmesan. Just take a carrot peeler and shave a block of the cheese. Shaved Parmesan is so much prettier than the grated version.

bold = *foods pollinated or produced by bees*

turkey roulade with cranberry chutney

Here's a fancy and different way of serving turkey. Don't worry; like all of my recipes, this one is quite doable and delicious. If you are tired of having a whole turkey for Thanksgiving, give this option a whirl. You could also use any ground meat, alone or in combination. Take your favorite meatloaf recipe and wrap it in the pastry as directed. Snazzy!

The ingredients:

- 1 sheet frozen puff pastry
- 1 pound turkey meat from turkey breast and/or thigh meat, cut into 1-inch chunks
- 3 eggs
- ½ cup chicken stock
- ¼ cup **Marsala**
- 2 tablespoons unsalted butter

    Unbleached all-purpose flour, for the pastry

- ½ cup **celery** sliced on the diagonal
- ½ cup sliced button mushrooms
- ½ cup **cranberry** chutney (page 164), plus more for serving

    Fresh **parsley** sprigs for garnish

Here's what you do:

1. Following the instructions on the package, thaw the puff pastry. This will take 30 to 45 minutes, depending on the type of pastry. You should be able to unfold it without breaking. Set aside.

2. Pulse the turkey in a food processor until it is the consistency of ground beef. Add 2 of the eggs, the chicken stock, the Marsala, and the butter. Pulse again briefly, until just combined.

3. Roll out the puff pastry on a floured surface until it is a 12- or 13-inch square. Cover a baking sheet with parchment paper and place the pastry on top.

4. Form the minced turkey mixture into a log and position it down the center of the prepared puff pastry. Make an indentation down the length of the turkey and place the celery, mushrooms, and cranberry chutney along the indentation. Cover the indentation with the meat, forming a log once again.

5. Wrap the puff pastry around the turkey, neatly folding the ends and top together, rolling or tucking the edges together, and pinching to seal any gaps. Make the pastry-covered log as round as possible, like a Yule log — try to avoid a flattened version, like a strudel.

6. Preheat the oven to 450°F.

7. Combine the remaining egg with 1 tablespoon water in a small bowl or cup. Stir well until completely mixed. Brush the roulade with the egg wash, being careful to brush every bit of the exposed pastry.

8. Bake for 10 minutes at 450°F, then reduce the heat to 375°F. Continue to bake 30 to 45 minutes longer, until the pastry has risen and is a toasty, golden color and the turkey has reached an internal temperature of 165°F. Generally speaking, once the pastry has cooked, the meat will be cooked too. Remove the roulade from the oven. Transfer it to a serving platter, using the parchment paper to help you. Allow to rest for about 10 minutes.

9. Slice and serve, garnishing with the parsley and accompanying with a bowl of extra cranberry chutney.

Serves 6

sweet fact

On average, each person in the United States eats about 1 pound of honey in a year — the life's work of about 800 bees.

hot mulled (sherried) apple cider

If you are having guests over when it is apple cider season, put a pot of this on the back burner of your stove about an hour before everyone arrives. By the time they get to your house, it will be filled with the aroma of the cider, the spices, and the oranges.

Having kids in for a sledding party? This is a great hot drink for the parents as the kids warm up with hot chocolate. If the kids want some hot cider, serve it to them in a cup with a whole stick of cinnamon, and show them how to use it as a straw. De-lish!

The ingredients:

- 1 gallon **apple cider**
- 1 **orange**, unpeeled, cut into slices
- ¼ cup whole cloves
- 4 sticks **cinnamon**
- ¼ cup **honey**, preferably cranberry honey
- 1 cup **sherry** (optional)

Here's what you do:

1. Combine the cider, orange slices, cloves, cinnamon, and honey in a large pot over medium heat. If you are picky about things floating in your cider, make a little bundle out of cheesecloth and place the cinnamon and cloves inside before adding to the cider. I like to chew on cloves, so I just toss everything in. Bring to a boil, and then reduce to a simmer over low heat for an hour or so to spread these lovely winter aromas around your home.

2. If you're serving it to adults, add the sherry, as my mother did. It might make everyone want to go sledding!

Serves 16

elsie's cranberry pie

This is another fun dessert. Though called a pie, it doesn't have a pesky crust, so calm your worries about that. My version is made with cranberry honey instead of sugar. If you need a quick dessert for a family supper or a potluck contribution, give this one a shot. Outside of cranberry season, use seasonal berries.

Serves 6

The ingredients:

- 3 cups fresh **cranberries**
- ½ cup chopped **pecans**
- ½ cup **honey**, preferably cranberry honey
- 2 eggs
- ¾ cup unbleached all-purpose flour
- ½ cup (1 stick) butter, melted

Here's what you do:

1. Preheat the oven to 325°F.

2. Butter a 10-inch pie plate and fill with the cranberries and pecans. Drizzle with ¼ cup of the honey.

3. Beat the eggs in a medium bowl until light lemony yellow. While still beating, add the remaining ¼ cup honey, the flour, and the melted butter, and mix thoroughly. Spread the mixture on top of the cranberries.

4. Bake for 45 minutes, or until the crust is a nicely toasted light brown. Let cool slightly, and serve while still warm. Leftovers (if any!) will be great for breakfast.

bold = *foods pollinated or produced by bees*

chestnut

How doth the little busy bee
Improve each shining hour,
And gather honey all the day
From every opening flower!

— ISAAC WATTS

Chestnut honey is one of the

stronger honeys. I was introduced to it on a trip to Tuscany, where chestnut honey is a prized local taste. Italians appreciate many more bitter flavors than we do, and it was a big taste explosion for my palate. The color, flavor, and smell of chestnut honey vary depending on the source of the chestnut trees. Descriptors go from light and slightly pungent to extremely strong, breathtaking, and lingering.

In this chapter, I am not recommending chestnut honey for all the recipes. It is simply too much for most people. However, if you love strong tastes and your guests are adventurous, give it a whirl. At every honey tasting I conduct, there is a good handful of tasters who love its strong, assertive flavor.

COLOR	SMELL	TASTE	AFTERTASTE
Dark orange	Pungent; farmlike	Very strong and distinctively sweet with a curious musty aroma that can be off-putting but really appeals to folks who enjoy strong tastes	Medicinal

pears and pecorino with chestnut honey

This is a very simple appetizer, but it completely stopped me in my tracks the first time I traveled to Tuscany. Pecorino cheese, made from sheep's milk, is very popular in Tuscany, though the cheese makers originally migrated there from Sardinia. American cheese makers are now making some wonderful pecorino cheeses right here, having spent time in Italy at the feet of those master cheese makers, but if you can't find really good pecorino, substitute aged cheddar. Avoid commercially available Pecorino Romano, as it has an unpleasantly strong taste. Look for Pecorino di Pienza instead.

The trick here is to encourage your guests to make sure that each bite includes a bit of pear, a bit of honey, and a bit of cheese. It is one of those wonderful combinations where the whole is far more interesting than each of the parts.

Serves 4
as an appetizer

The ingredients:

- 2 ripe but firm **pears**
- ½ pound aged pecorino cheese
- 2 tablespoons **honey**, preferably chestnut honey or a local, dark honey

Here's what you do:

1. Cut the pears in half from the stem to the blossom end. Remove the core and cut each half into 5 long wedges.

2. Slice the cheese into thin triangles.

3. Place the sliced pears on a platter and put a slice of cheese on each pear slice. Dip a fork into the honey and drizzle a thin stream over the fruit and cheese slices.

4. Serve immediately!

cream of chestnut soup

My mother often kept my sisters and me occupied around Thanksgiving time by giving us chestnuts to peel. When I moved to New York City after college, a bag of hot chestnuts would keep my pockets filled with fragrant warmth. You may want to peel your own chestnuts, which is easy enough to do, though kind of timeconsuming, or you could purchase whole, peeled chestnuts. Look for them at your local gourmet grocery store in the winter. This is a delicious soup that's easy, comforting, and reminiscent of the wintry old days.

Chestnut honey has too strong a flavor for this recipe. Instead, I suggest eucalyptus, a dark honey that doesn't have the overly assertive taste of chestnut.

The ingredients:

- 2 tablespoons butter
- 1 large **onion**, peeled and cut into large chunks
- 2 pounds cooked, peeled **chestnuts**
- 2 large **carrots**, peeled and cut into ½-inch slices
- 4 cups chicken or vegetable stock, plus more if needed
- 1½ cups heavy cream, plus more if needed
- 2 tablespoons **honey**, preferably eucalyptus or other dark honey
- ½ cup dry **sherry**
- 1 teaspoon kosher salt

    Freshly ground black pepper

Here's what you do:

1. Melt the butter in a large soup pot over low heat. Add the onion and sauté until transparent, about 5 minutes. Add the chestnuts, carrots, and stock. Simmer over low heat until the chestnuts are very tender (until you can poke a fork through one), about 10 minutes. Remove from the heat.

2. Blend the soup with an immersion blender until completely smooth, or drain the vegetables in a colander and pulse them in a food processor until smooth; return to the pot and blend with the broth.

3. Add the cream, honey, and sherry. Add the salt and a few grinds of pepper. Taste and adjust the seasonings. If the soup is too thick, add additional stock or cream. Warm over medium-low heat but do not boil, as you do not want a curdled mess on your hands.

Serves 6–8

pears with blue cheese, toasted pecans, and chestnut honey vinaigrette

I get excited by the proliferation of pears in the market in the winter. I imagine what it would be like to live in Washington or Oregon. And so, though they are not local to me in December, pears are available and abundant and become the foundation for this delicious salad.

Sweet, salty, bitter, and sour: All four tastes are in this salad, which makes it a memorable one for your guests.

The ingredients:

FOR THE VINAIGRETTE

- 2 tablespoons **orange juice**
- 2 tablespoons **red wine vinegar**
- 1 tablespoon **honey**, preferably chestnut honey
- ¼ cup extra-virgin olive oil

  Sea salt

  Freshly ground black pepper

FOR THE SALAD

- ¼ cup **pecan** pieces, toasted
- 4 ripe but firm **pears**
- ¼ cup crumbled blue cheese (I like Maytag)

Serves 8

Here's what you do:

1. To make the vinaigrette, combine the orange juice, vinegar, and honey in a small bowl and stir with a wire whisk until well mixed. Drizzle the oil into the bowl in a thin stream, whisking constantly until well blended. This will take 2 to 3 minutes. Season with salt and pepper to taste.

2. To make the salad, toast the pecans in a small dry saucepan for 2 to 3 minutes over medium heat, watching carefully and tossing often so they don't burn.

3. Cut the pears in half from the stem to the blossom end. Remove the core, and cut each half in half again.

4. Arrange the pear quarters on individual salad plates. Sprinkle with the cheese and toasted pecans and, just before serving, drizzle with the vinaigrette.

bold = *foods pollinated or produced by bees*

winter fruit–stuffed pork tenderloin

I really like fruit and pork. And I really like blue cheese. This is my version of a complicated recipe that is usually made with a pork loin. It's much faster to put together with the smaller tenderloin. I suggest an easy honey glaze, which gives it a lovely, mildly sweet finish.

I recommend using something other than chestnut honey in this recipe, unless you are really fond of its assertive flavor. Try sourwood honey for a lightly spiced finish.

The ingredients:

½ cup dried **apricots**

½ cup dried **cherries**

½ cup dried **figs**

2 pork tenderloins (about 3 pounds total)

1 teaspoon salt

1 teaspoon freshly ground black pepper

1 cup crumbled blue cheese

2 tablespoons extra-virgin olive oil

4 tablespoons butter

¼ cup **honey**, preferably sourwood honey

Here's what you do:

1. Mince the apricots, cherries, and figs by hand or in a food processor.

2. Slice the tenderloins lengthwise, almost all the way through. Open them up and lay them flat. Place each tenderloin on a large piece of plastic wrap. Cover with another piece of plastic wrap and pound each piece of meat with a meat tenderizer until it is about ½ inch thick. Remove the top piece of plastic.

3. Season the surface of the pork with the salt and pepper. Divide the fruit mixture in half and spread evenly on the cut surface of each tenderloin. Top each with half of the cheese. Roll up each tenderloin, using the bottom piece of plastic to help you, tucking in the fruit and cheese as you go. Tie kitchen string every 2 inches around the tenderloins, continuing to push in any fruit or cheese that may fall out.

recipe continues on following page >

---

bold = *foods pollinated or produced by bees*

4. Preheat the oven to 350°F.

5. Heat the oil in a large skillet over medium-high heat. Sear the tied tenderloins, turning as each side is browned. Be careful when searing the open side, as some fruit and cheese might fall out. You are just trying to seal in the meat juices, not trying to cook the pork all the way through.

6. Combine the butter and honey in a microwaveable bowl and microwave on high for about 20 seconds, or until the butter is melted. Drizzle the butter over the tenderloins.

7. Place the tenderloins on a baking sheet. Bake for 15 to 20 minutes, or until the meat reaches an internal temperature of 150°F. Remove the baking sheet and allow the tenderloins to sit for at least 10 minutes before slicing. This will keep the juices in the meat rather than all over your kitchen counter.

8. Snip off and discard the strings. Slice the pork into 1-inch-thick pieces and serve.

Serves 8

you've been warned!

The job of the guard bee is to keep unwelcome visitors out of the hive. A beekeeper or a passerby whom the guard bees regard as threatening gets warned with a bump and, if he or she does not leave, gets stung. If you're around a hive of honeybees and one bumps into you more than once, scoot! After three bumps, you will probably be stung!

bavarian cabbage

When I first moved to Asheville, we had an interesting restaurant owned by an Indian man and his midwestern American wife. The husband made the most wonderful food, some Indian and some not, including a German-style sweet-and-sour cabbage that accompanied all of his wife's rich, American main-course offerings. I would have been satisfied with a plate full of his cabbage, though her roasts, stews, and schnitzels were hard to resist. Here's my version of their cabbage, sweetened with honey, of course. This is a terrific accompaniment to the pork tenderloin (page 183). Make the cabbage ahead of time. It will continue to get better over the course of a few days.

Chestnut honey is too assertive for this dish, so substitute a milder honey, like orange blossom or tupelo.

The ingredients:

- 6 slices applewood-smoked bacon
- 8 cups thickly sliced **red cabbage**
- 4 cups tart **apples**, unpeeled, cut into chunks
- ½ cup **apple cider vinegar**
- ⅓ cup **honey**, preferably orange blossom or tupelo or another mild honey
- Sea salt
- Freshly ground black pepper
- ¼ cup sour cream or Greek yogurt (optional)

Here's what you do:

1. Cook the bacon in a large soup pot over low heat until the bacon drippings collect in the bottom of the pot. Transfer the bacon to paper towels to drain and cool. Cut into pieces.

2. Add the cabbage, apples, vinegar, honey, ½ cup water, and some salt and pepper to the pot with the bacon drippings. Cook slowly over very low heat, covered, until the cabbage is tender, about 35 minutes, checking and stirring on a regular basis. Taste and add more salt and pepper if needed.

3. You may serve this as is, or with the addition of a dollop of sour cream. Garnish with the bacon pieces.

Serves 6–8

Hold the Bacon

If making a vegetarian version, use 4 tablespoons of olive oil instead of the bacon drippings, though you will have to add more salt in the final seasoning.

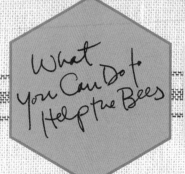

What you Can Do to Help the Bees

There are small, yet meaningful, things you can do to help honeybees. Elements of this list are provided with permission by the producers of the documentary film *Queen of the Sun: What Are the Bees Telling Us?* and by Collective Eye, Inc.

Don't use pesticides. It might take longer, but weeding by hand is one of the most considerate actions you can take to help the bees. Honeybees collect pollen and nectar from all flowers, and ingesting poisoned plants poisons the bees. Remember, bees are insects, and pesticides and herbicides affect more than just mosquitoes and crabgrass.

Buy local produce. Pick up produce from your farmers' market or from the organic section of your produce department. Choose farmers who take the extra step to farm without pesticides.

Buy honey from a local beekeeper. This is one of the easiest things to do. Shop at your local farmers' market and ask the beekeeper how he or she keeps the bees. There is a growing movement to eliminate chemical treatments. As more and more consumers vote with their wallets, beekeepers will pay attention.

Plant a bee garden. Bees need nectar, so if you have room, plant flowers for them. Native perennials are great; ask your local nursery what grows best in your area. Also, bees like to visit just one kind of flower during each foraging flight, so plant flowers in large groups if you can.

Make water available. Bees need to drink and will visit a bird feeder or a small pond (or a pool!). Keep clean water in your yard and provide something for them to stand on, too. A little pile of rocks or floating water plants will give them a safe place to rest while they sip. I have borders of wine bottles around some of my flower beds. Creating them was a simple matter of pounding the stem end of empty bottles into the ground. Each bottle collects water in its inverted base, which I've filled with pebbles so the bees won't drown. Now the bees visit my borders, stand on the pebbles, and drink — an adorable sight!

Don't kill honeybees. This goes without saying. Spread the word that honeybees are good. If you are fortunate enough to find a swarm, call your local beekeeping group. Someone will capture the swarm, even if it's in the wall of your home. And if you do have a problem with yellowjackets or wasps or other noxious insects, tell your exterminator that you do not want to use poisons. They have alternative treatments.

Leave an area on your property wild, if you can. Weeds are just plants growing where we don't want them, but they can offer sanctuary and sustenance to honeybees.

Become a beekeeper yourself. It is a magical hobby, an inspirational pastime, and a life-changing experience.

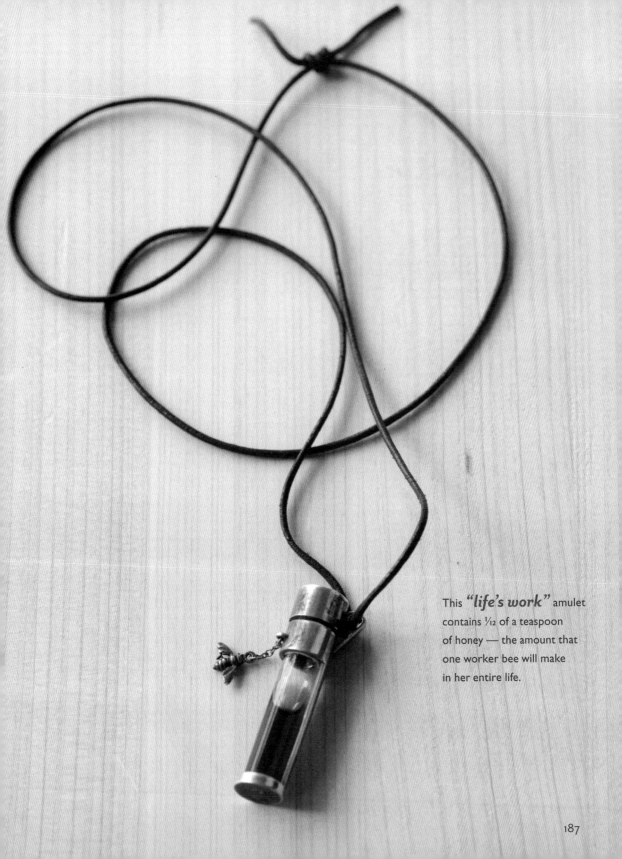

This **"life's work"** amulet contains 1/12 of a teaspoon of honey — the amount that one worker bee will make in her entire life.

poached pears with crème anglaise

This is such a delicious and easy dessert. You can slice and prepare the pears before your guests arrive, turn on the heat as you sit down to dinner, and have dessert ready as you are clearing dinner plates. You'll need to make the crème anglaise earlier, but that's quite doable. Use a light, unassertive honey in this recipe. Chestnut honey will overpower the fruit.

The ingredients:

FOR THE PEARS
- 8 firm **pears**, such as Bartlett
- 2 quarts **apple cider**
- 1 stick **cinnamon**
- 1 tablespoon whole cloves

FOR THE CRÈME ANGLAISE
- 1 cup heavy cream
- 2 teaspoons vanilla extract
- 4 egg yolks
- ⅓ cup **honey**, preferably sage or acacia or another light honey

Here's what you do:

1. To make the pears, peel them but leave the stem on. Cut a thin slice off the bottom of each pear so that it can stand up. Place in a medium deep saucepan, one just big enough to hold all the pears, if possible. Pour the apple cider over the pears, covering them completely. Add the cinnamon and the cloves. **NOTE**: You can do this ahead of time if you like, leaving the pears in the saucepan, unheated, on your stove until you are ready to cook them.

2. Heat the pears in their cider bath over medium-low heat until the cider is simmering. Turn the heat down to low so that the liquid barely bubbles — a minimal simmer, so to speak. Cook at this temperature for about 30 minutes. Turn off the heat and let the pears sit in the poaching liquid until ready to serve.

3. To make the crème anglaise, heat the cream and vanilla in a small saucepan over low heat until bubbles begin to form around the edges of the pan. Watch carefully! You do not want the mixture to boil, as it will curdle. Remove from the heat and allow to cool slightly.

4. Whisk the egg yolks with the honey in a medium bowl.

5. Very slowly pour half of the heated cream into the egg yolk mixture, whisking constantly. Then mix this egg-cream mixture back into the remaining warm cream in the saucepan. Return to the stove and slowly warm over very low heat, stirring constantly, until the mixture coats the back of a spoon. Remove from the heat and set aside until ready to use. Again, be very attentive here so the mixture does not boil!

6. Place a spoonful of the crème anglaise on each plate and position a whole pear on top. Serve with a bowl of the leftover crème anglaise — your guests will probably want more.

Serves 8

"guggeluh muggeluh"
(a warmed honey drink)

When he was a little boy, my sweet friend Ken's mother, Rose, used to lull him to sleep with this simple warm drink. A Yiddish dictionary lists it as "Guggle Muggle," but my moniker is much more Yiddish-y, I think, and, besides, that's what Ken's mother called it. It's an easy thing to make.

The ingredients:

- 4 cups milk
- 4 tablespoons **honey**, preferably sourwood or another mild honey
- 4 tablespoons butter (optional)

Here's what you do:

1. Add the milk to a medium saucepan and warm over low heat. Whisk in the honey.

2. Pour into mugs. If it is a very cold night, add 1 tablespoon butter to each mug to help keep everyone warm all night long. Enjoy!

Serves 4

sweet fact

In one nectar-foraging trip, a honeybee visits 50 to 100 flowers, all of one kind.

Good night, Rose.

Good night, Kenny.

Good night, Bees.

recipes organized by course

salads

main courses

sides

desserts

beverages

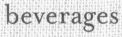

foods pollinated by honeybees

FRUITS

apples
apricots
blackberries
blueberries
boysenberries
cantaloupe
carambola
cherimoya
cherries
citron
coconut
cranberries
currants
dewberries
elderberries
gooseberries
grapefruit
grapes
guava
honeydew
huckleberry
jujube
kiwifruit
lemons
limes
loquats
lychee
mangos
muskmelons
nectarines
oranges
papayas
passion fruit
peaches
pears
persimmons
plums
pomegranates
pomelos
quince
raspberries
sapote
serviceberries
strawberries
tamarind
watermelon

VEGETABLES

artichokes
asparagus
avocados
beets
black-eyed peas
broccoli
Brussels sprouts
cabbages
carrots
cauliflower
celery
chayote
chicory
Chinese cabbage
collards
cow beans
cucumbers
eggplant
garlic
gourds
kale
kohlrabi
leeks
lima beans
mustard
okra
onions
parsley
peppers
pimientos
pumpkin
radishes
rutabaga
scarlet runner
  beans
squashes
turnips
zucchini

NUTS & SEEDS

almonds
cacao
cashews
chestnuts
coffee
flax
hazelnuts
kola nuts
macadamia nuts
sesame
walnuts

GRAINS

alfalfa
buckwheat
clover
vetch

HERBS & SPICES

allspice
anise
caraway
cardamom
chives
cinnamon
coriander
dill
fennel
lavender
mint
mustard
nutmeg
oregano

OIL CROPS

canola
cottonseed
palm seed

where to find honey varietals

For starters, check the honey section of your local fancy food or higher-end health food store. You may very well be able to find some of the honeys mentioned in this book.

If you live in an area with a good farmers' market, ask if the beekeeper has anything that resembles one of the honeys you are looking for. Many beekeepers are knowledgeable on the tastes of their own honey and can help you out. Sample their honey, compare it to the flavor description in this book of the one you want, and come as close as you can.

A great source for honey online is the National Honey Board's "honey locator" (www.honeylocator.com).

The Savannah Bee Company is a fine source of honey and a fun store to visit. If you are in the Southeast, visit one of their stores, either in Savannah or in Charleston. If you're not able to visit, just check out their website (www.savannahbee.com).

Listed below are all the honeys I mention and where I found them. A simple Google search will also be helpful.

Acacia
Savannah Bee Company
Savannah, Georgia
800-955-5080
www.savannahbee.com

Avocado
Bennett's Honey Farm
Fillmore, California
805-521-1375
www.bennetthoney.com

Blueberry
New England Cranberry Co.
Lynn, Massachusetts
800-410-2892
http://newenglandcranberry.com

Chestnut
Chestnut honey, the kind I like, comes from Tuscany. Bees collect nectar from trees near Mount Amiata, the tallest mountain in Tuscany. If you can't find it in your gourmet store, check online or, better yet, go to Italy to get some yourself!

Cranberry
New England Cranberry Co.
(See blueberry)

Eucalyptus
Bennett's Honey Farm
(See avocado)

Orange Blossom
Fruitwood Orchards Honey Farm
Monroeville, New Jersey
856-881-7748
http://fruitwoodorchardshoney.com

Raspberry
Fruitwood Orchards Honey Farm
(See orange blossom)

Sage
Honey Pacifica
Long Beach, California
562-938-9706
www.honeypacifica.com

Sourwood
Wild Mountain Apiaries & Beekeeping Supply
Asheville, North Carolina
828-484-9446
www.wildmountainbees.com

Tulip Poplar
Wild Mountain Apiaries & Beekeeping Supply
(See sourwood)

Tupelo
Savannah Bee Company
(See acacia)

suggested reading

Benjamin, Alison and Brian McCallum. *Keeping Bees and Making Honey*. David & Charles, 2008.

Bishop, Holley. *Robbing the Bees*. Free Press, 2005.

Bonney, Richard E. *Storey's Guide to Keeping Honey Bees*. Storey Publishing, 2010.

English, Ashley. *Keeping Bees with Ashley English*. Lark Crafts, 2011.

Flottum, Kim. *The Backyard Beekeeper*, rev. ed. Quarry Books, 2010.

———. *The Backyard Beekeeper's Honey Handbook*. Quarry Books, 2009.

Fisher, Rose-Lynn. *Bee*. Princeton Architectural Press, 2010.

Hubbell, Sue. *A Book of Bees*. Mariner Books, 1988.

Jacobsen, Rowan. *Fruitless Fall*. Bloomsbury USA, 2008.

Morrison, Alethea. *Homegrown Honey Bees*. Storey Publishing, 2013.

Ryde, Joanna. *Beekeeping*. Skyhorse Publishing, 2010.

Schacker, Michael. *A Spring without Bees*. Lyons Press, 2008.

Seeley, Thomas D. *Honeybee Democracy*. Princeton University Press, 2010.

Steiner, Rudolf. *Bees*. Anthroposophic Press 1998.

Unstead, Sue. *The Beautiful Bee Book*. School Specialty Publishing, 2006.

metric conversions

Unless you have finely calibrated measuring equipment, conversions between U.S. and metric measurements will be somewhat inexact. It's important to convert the measurements for all of the ingredients in a recipe to maintain the same proportions as the original.

GENERAL FORMULA FOR METRIC CONVERSION

Ounces to grams	multiply ounces by 28.35
Grams to ounces	multiply grams by 0.035
Pounds to grams	multiply pounds by 453.5
Pounds to kilograms	multiply pounds by 0.45
Cups to liters	multiply cups by 0.24
Fahrenheit to Celsius	subtract 32 from Fahrenheit temperature, multiply by 5, then divide by 9
Celsius to Fahrenheit	multiply Celsius temperature by 9, divide by 5, then add 32

APPROXIMATE EQUIVALENTS BY WEIGHT

U.S.	METRIC
¼ ounce	7 grams
½ ounce	14 grams
1 ounce	28 grams
1¼ ounces	35 grams
1½ ounces	40 grams
2½ ounces	70 grams
4 ounces	112 grams
5 ounces	140 grams
8 ounces	228 grams
10 ounces	280 grams
15 ounces	425 grams
16 ounces (1 pound)	454 grams
0.035 ounces	1 gram
1.75 ounces	50 grams
3.5 ounces	100 grams
8.75 ounces	250 grams
1.1 pounds	500 grams
2.2 pounds	1 kilogram

APPROXIMATE EQUIVALENTS BY VOLUME

U.S.	METRIC
1 teaspoon	5 milliliters
1 tablespoon	15 milliliters
¼ cup	60 milliliters
½ cup	120 milliliters
1 cup	230 milliliters
1¼ cups	300 milliliters
1½ cups	360 milliliters
2 cups	460 milliliters
2½ cups	600 milliliters
3 cups	700 milliliters
4 cups (1 quart)	0.95 liter
1.06 quarts	1 liter
4 quarts (1 gallon)	3.8 liters

index

Page numbers in *italic* indicate photos; those in bold indicate charts.

sweet fact

Queens are painted with a bright dot, the color of which lets the beekeeper know when she joined the colony. A yellow dot, for instance, indicates that the queen was added in a year ending with a 2 or a 7.

other storey titles you will enjoy

brewing made easy
2nd edition, by Joe Fisher and Dennis Fisher.
A foolproof starters' guide to brewing great beer at home — includes step-by-step
instructions and 25 recipes.
96 pages. Paper. ISBN 978-1-61212-138-3.

the fresh egg cookbook
by Jennifer Trainer Thompson.
A wealth of delicious recipes for using eggs from local farms and your own backyard.
192 pages. Paper. ISBN 978-1-60342-978-8.

homegrown honey bees
by Alethea Morrison. Photography by Mars Vilaubi.
A beginner's guide to the first year of beekeeping, from hiving to harvest.
160 pages. Paper. ISBN 978-1-60342-994-8.

hot sauce!
by Jennifer Trainer Thompson.
More than 30 recipes to make your own, plus 60 more recipes for cooking with
homemade or commercial sauces.
192 pages. Paper. ISBN 978-1-60342-816-3.

maple sugar
by Tim Herd.
From sap to syrup: the history, lore, and how-to behind this sweet treat.
144 pages. Paper. ISBN 978-1-60342-735-7.

raw energy
by Stephanie Tourles.
More than 100 recipes for delicious raw snacks: unprocessed, uncooked, simple, and pure.
272 pages. Paper. ISBN 978-1-60342-467-7.

**These and other books from Storey Publishing are available
wherever quality books are sold or by calling 1-800-441-5700.
Visit us at www.storey.com or sign up for our newsletter
at www.storey.com/signup.**